News from the past:
Progress in African archaeobotany

Proceedings of the 7th International Workshop on African Archaeobotany in Vienna, 2 – 5 July 2012

Advances in Archaeobotany

Volume 3

Contact the Editorial Board

contact@archaeobotany.eu
www.archaeobotany.eu

ISSN

2405-7452

Contact the Publisher

Barkhuis
Kooiweg 38
9761 GL Eelde
the Netherlands
info@barkhuis.nl
www.barkhuis.nl

News from the past:

Progress in African archaeobotany

Proceedings of the 7th International Workshop on African Archaeobotany in Vienna, 2 – 5 July 2012

Edited by

Ursula Thanheiser

Barkhuis Publishing
Groningen 2016

Book design: Barkhuis Publishing, Groningen
Cover design: Nynke Tiekstra, Coltsfootmedia, Rotterdam
Copy editor: Suzanne Needs-Howarth, Toronto

Cover image: Ababda nomad weeding a field of Sorghum (*Sorghum bicolor*) in the coastal area of the Red Sea by using a hoe. Because the amount of available water is limited, the distance between the individual crop plants is relatively large so that weeding during the growing season is possible. The roots of the wild plants are cut by shallow hoeing, which reduces the competition for water and minerals (Areb Saleh, Egypt; January 1997; R.T.J. Cappers). From the *Digital atlas of traditional agricultural practices and food processing*, R.T.J. Cappers, R. Neef, R.M. Bekker, F. Fantone, & Y. Okur

universität wien

The publication was subsidized by the Faculty of Historical and Cultural Studies of the University of Vienna

ISBN 9789492444028

Table of contents

Preface

The seventh International Workshop on African Archaeobotany was held in Vienna, 2–5 July 2012. It was organized under the auspices of Vienna University by the archaeobotany section of the Vienna Institute for Archaeological Science (VIAS), and it enjoyed the hospitality of the Department of African Studies. We are grateful for the logistical support we received from the university, the Department of African Studies, and the administration of VIAS.

Sincere thanks are due to copy editor Suzanne Needs. Without her endless patience this book would never have reached completion. The production of this volume was subsidized by the University of Vienna.

The International Workshop on African Archaeobotany was established by Krystyna Wasylikowa, and its first meeting took place in 1994, in Mogilany, Poland. At a time when archaeobotany was still in its infancy for most of the African continent, IWAA provided a platform for presenting ideas and discussing results. It was also a way of getting practical help with obtaining reference material and with the identification of taxa usually unknown to archaeobotanists trained in more northern countries. Since then, IWAA has taken place every three years, and peer-reviewed proceedings have been published for each of these meetings. Previous IWAA meetings were organised by Marijke Van der Veen in Leicester (1997), Katharina Neumann in Frankfurt am Main (2000), René Cappers in Groningen (2003), Dorian Fuller and Mary Anne Murray in London (2006), and Ahmed Gamal el-Din Fahmy in Cairo (2009).

Since its inception two decades ago, IWAA has developed into a tightly knit community of scholars from all continents who share a profound interest in African ways of plant exploitation, trade networks, questions of origin, domestication and subsequent dispersal of African crops, as well as the introduction of crops of Asian and American origin. The workshop is blessed with easy communication and informal sharing of results and has always encouraged young colleagues to present their results and test their ideas. During IWAA's 20-year history, the range of research presented by participants has widened considerably. Having initially dealt mainly with plant macroremains, IWAA now includes charcoal, pollen, and phytoliths, as well as aDNA, stable isotope, ethnographic, historic, and linguistic studies.

Most of the papers in this volume were presented at the 2012 workshop in Vienna. They address past interrelationships between people and plants as evident in the rich archaeobotanical, ethnographic, and linguistic record of Africa.

Ursula Thanheiser

Ahmed Gamal el-Din Fahmy

1962–2013

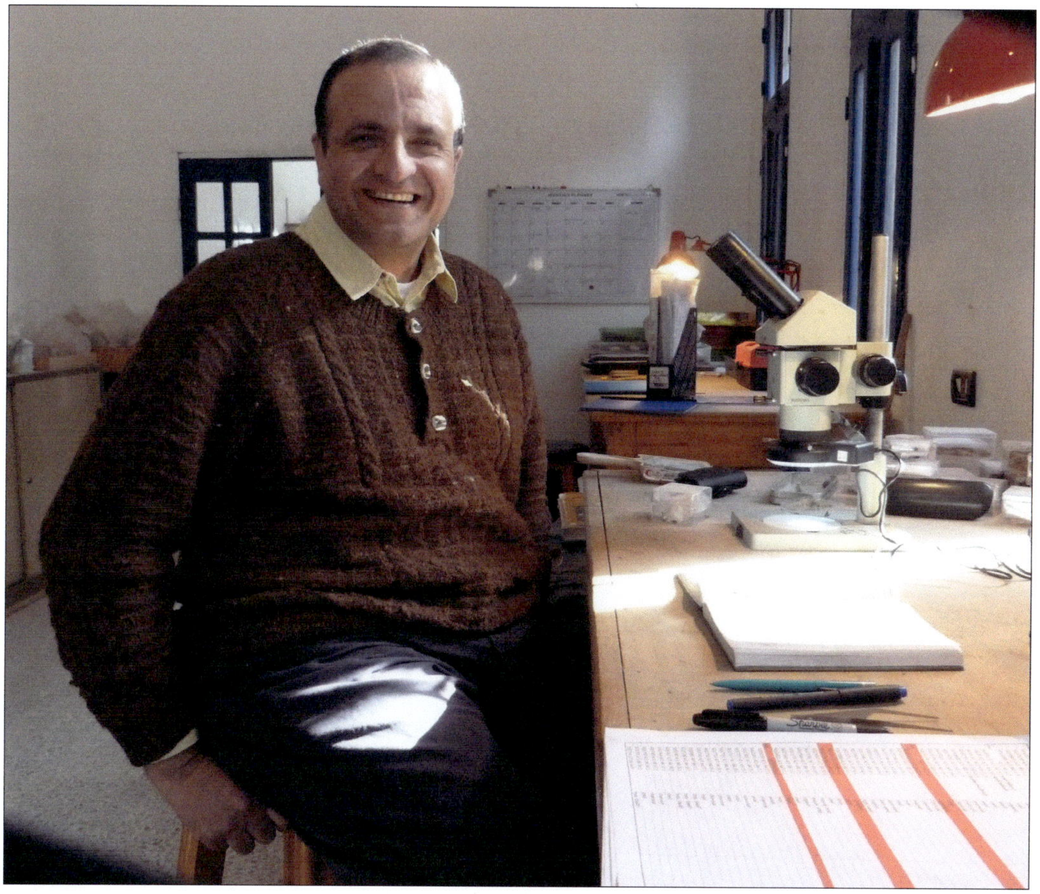

Ahmed Fahmy in the field lab at Helwan, December 2011.

The archaeobotanical world lost one of its great proponents with the sad demise of Dr Ahmed Gamal el-Din Fahmy, Professor of Botany at Helwan University, Egypt. We have also lost a dear friend.

Ahmed, more than most, steadily took the study of ancient Egyptian plant remains into the twenty-first century with his innovative and motivational approach. Any archaeobotany project that Ahmed applied himself to would shine due to his hard work, creative scientific approach to his area of research, fundamental command of the subject, ground-breaking approach, and thoughtful integration of data from other specialties to highlight the many facets of the discipline.

I first met Ahmed in 1987, at the Herbarium at Cairo University. We always joked that this was back when he was a 'baby botanist'. As a true botanist, though, he was very interested in archaeobotany even then, and he would often come to where I was working on a dig at Memphis, south of Cairo, for 'microscope time' on the ancient equivalents of the plants he knew so well. Memphis is one of the richest sites I've worked on in Egypt, so for both of us there was much to learn in modern and ancient botany. It was great spending time in the lab with Ahmed.

Ahmed went on to excel in the field of archaeobotany, and he leaves a legacy of vital contributions to Egyptian archaeobotany. But more importantly, he was an exceptional person too—and he remained a great friend till the end.

Rest in peace.

Mary Anne Murray

The history of conifers in Egypt, part I: Mediterranean cypress (*Cupressus sempervirens* L., Cupressaceae)

Victoria Asensi Amorós

Xylodata SARL
4, Place Violet
75015 Paris, France

E-mail: victoria.asensi@free.fr

Abstract. The article describes archaeological remains of *Cupressus* sp. and *Cupressus sempervirens* found in Egypt and discusses their wider environment, namely, the Middle East and the Mediterranean Basin, both regions accessible to the ancient Egyptians. The physical appearance of cypress wood, its anatomy, and its current and past uses are also reviewed. Archaeological data concerning made objects from the Predynastic period until the Islamic era are presented. Finally, the article also briefly mentions Egyptian iconography and epigraphy relating to cypress.

Keywords: cypress, wood anatomy, utilisation, Egypt

1. Introduction

While the current Egyptian indigenous flora comprises just a single species of conifer, *Juniperus phoenicea* L., which is present in the Sinai (Täckolm 1974: 50), several species of conifers have been identified from the archaeological record (i.e. fir, *Picea*; cedar, *Cedrus*; and pine, *Pinus*). It therefore seems helpful to launch a series of papers discussing these different species of conifers and presenting the current state of research into their roles in past Egyptian society. I am arbitrarily starting with the genus *Cupressus* (cypress) and the species *C. sempervirens* L., being the only *Cupressus* species that has been identified from archaeological contexts in Egypt.

In this paper, I will deal with archaeological objects only, and not with charcoal or other remains. I will start with an overview of the distribution and ecology of cypress, as well as macroscopic and microscopic descriptions of the timber and its uses. Next I will discuss published identifications (including my own unpublished ones), in the form of a chronological catalogue. Epigraphic and iconographic data are also included. Unless otherwise stated, the listed identifications refer to *Cupressus* sp.

2. Distribution, Ecology, and Description

The genus *Cupressus* is represented by around 20 species distributed throughout the temperate zone in North America, Asia, and the Mediterranean region. The common species in the Mediterranean basin is *Cupressus sempervirens,* with the variety *pyramidalis* distributed as far as Asia Minor. The large-crowned variety (var. *horizontalis*), which is frequently admixed to stands of the var. *pyramidalis*, grows in Cyrenaica (eastern Libya), Greece, Cyprus, Rhodes,[1] Turkey, Syria, Iran, and Iraq.[2] In Syria and in Lebanon, this species is present in some localities, alone or in association with *Pinus brutia* Ten. (Turkish

1 In Cyprus, this species is very common in the dense forests in the mountains in the north of the island; in the Aegean islands, it appears in Kos, Rhodes, Karpathos, Samos, and Crete. In Crete, it is very well represented in the association *Cupressus-Acer orientalis*, up to an altitude of 1600 m (Paraskevopolou 1991: 195).

2 Being a characteristic tree of the forest and maquis, *Cupressus sempervirens* var. *horizontalis* is distributed as far south as Iran, in the mixed forests of *Acer monspessulanum* L., *Quercus iberica* M. Bieb., *Crataegus monogyna* Jacq., *Cerasus microcarpa* Bois., and *Carpinus orientalis* Mill. In Turkey, this species is confined to south-west Anatolia, where it is found along with *Pinus brutia* Ten. and *P. nigra* J. F. Arnold or in forests of *Cedrus* (Sherif & el-Taife 1986: 17; Zohary 1973: 347).

pine). In Palestine, there is a small patch of cypress forest.[3] The discovery of fossil cypress pollen in the mountains of Judea seems to suggest a wider distribution of these species in the past (Zohary 1973: 347).

From a palaeobotanical point of view, the original range of distribution of the Mediterranean cypress is still a matter of discussion. Certain scholars argue that its origin may be in Asia and that it disseminated from there during the Tertian era, eventually encompassing the whole Mediterranean area (Meiggs 1985: 46).

Cypresses are in general medium-sized trees and shrubs, but some individual trees can reach up to 30–40 m in height. *Cupressus sempervirens* may reach heights of 20–30 m and diameters up to 0.80 m at the base of the trunk. The trunk is frequently fluted, twisted, and knotty (Collardet & Besset 1988: 210). The leaves are small and opposite, generally with a longitudinal groove on the external face and with a blunt tip. The leaves often bear a dorsal resinous gland. The inflorescence is monoecious; the brown female cones are globose, subglobose, or ovoid, ranging from 2 to 4 cm in diameter (Sherif & el-Taife 1986: 16).

A remarkable species of cypress was discovered in 1924 in the Tassili Mountains in the Sahara (Camus 1926: 39). *Cupressus dupreziana* Camus is very closely related to *C. sempervirens*, but with a subdistichous branching pattern and small cones. This species is drought-resistant and grows very quickly in fresh soils. Another species that is highly resistant to drought is *C. atlantica* Gaussen. It is very common in the Moroccan forest (Callen 1976: 294, 310).

2.1 Cypress wood

Cypress wood is yellow-brown, sometimes tinged with pink. The sapwood is whitish, not very distinct, and narrow. The transition between early wood and late wood is gradual. The grain is straight to irregular, its texture fine and even. Resin canals may be present, appearing as occasional brown streaks. Cypress wood is not very resinous, and it is less odorous than cedar wood. It is light and tender, with little to moderate volume shrinkage. The wood is of excellent quality and offers good natural durability even though its mechanical strength is low to medium. It is easily worked and polished, resists decay in water, and is not attacked by insects.

Nowadays, because of its durability and water resistance, it is employed in construction, in exterior and interior cabinet work, as well as in situations where the wood is in direct contact with the ground. It is also used for furniture and decoration, including wood turning. Because of its very good acoustic properties, it is also used for the bodies of string-instruments (Collardet & Besset 1988: 210; Garcia Esteban et al. 1996: 184).

2.2 Cypress wood anatomy

The growth rings are distinct, with a gradual transition from early to late wood (Fig. 1). Tracheids with uniseriate (rarely biseriate) bordered pits are confined to the radial and the tangential walls in late wood. The pit apertures in early wood are rounded to slightly elliptical. The axial parenchyma is scarce, diffuse to banded, with smooth to faintly nodular end walls.

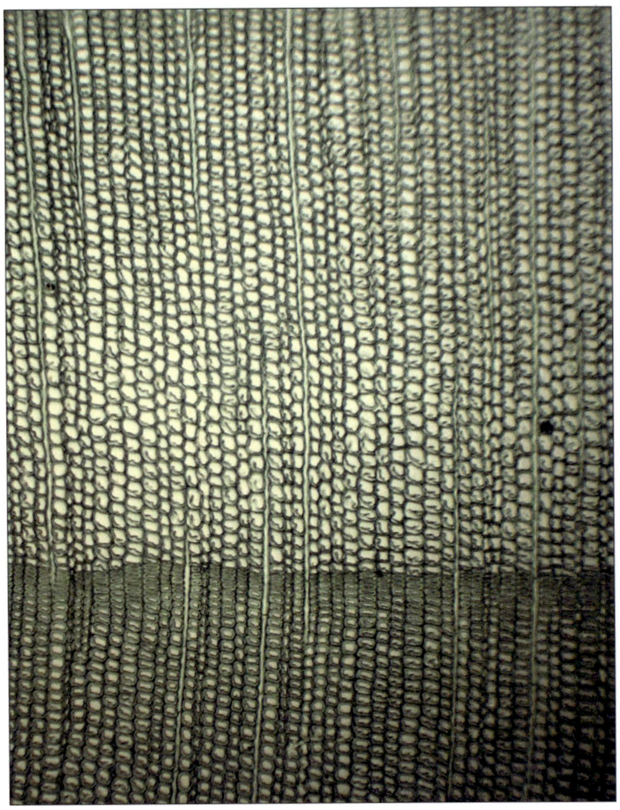

Figure 1. Transverse section, *Cupressus sempervirens*.

The rays are exclusively uniseriate, only rarely biseriate, 1–20 (30) cells high. Ray parenchyma cells have thin and smooth end walls (Fig. 2). Between 2 and 4 pits are found per cross field; these are cupressoid, with elliptical to slit-like apertures (Fig. 3). Resin ducts are absent.

The wood anatomy of *Cupressus* is very similar to that of *Juniperus* sp., but ray height and frequency (smaller in height, but more numerous in *Juniperus* sp.) and tangen-

3 Especially Gilead, in the hills of Edom.

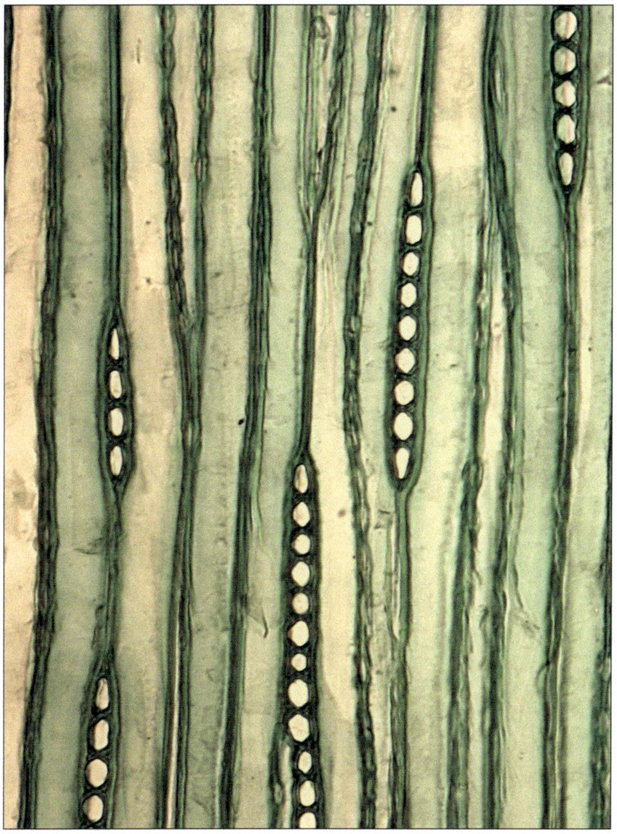

Figure 2. Tangential section, *Cupressus sempervirens.*

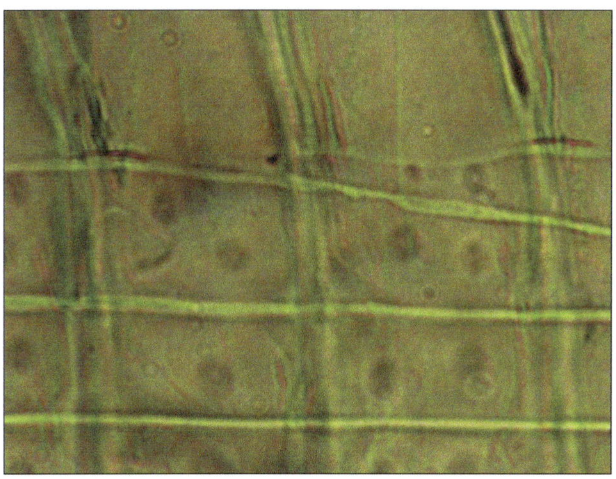

Figure 3. Radial section, *Cupressus sempervirens.*

tial cell walls (nodular in *Juniperus* sp., smooth or rarely nodular in *Cupressus* sp.) can help in distinguishing between these two genera.

2.3 Archaeological Data

See Appendix 1 (at the end of this chapter) for an overview of archaeological finds of *Cupressus* wood in Egypt.

3. Epigraphy and iconography

Charpentier (1981: paragraph 268) mentions this species under the generic term 'conifer' (*š*). It is acknowledged that, in the current state of research, the meanings of the different hieroglyphic names of the conifers are uncertain. However, for Bardinet, the word *mrw* (which is commonly thought to designate the cedar) would be the name for the cypress (for the cedar, he proposes the terms *snḏm* and *ssnḏm*). His arguments are: the red colour of *mrw* wood, its high frequency in shipbuilding, and the fact that it is a tree from the west and from Lebanon (Bardinet 2008: 73-84). It is true that the identifications made don't allow, in general, confirmation or rebuttal of the use of *Cupressus sempervirens*, or of another species. We must not forget that the softwood species most widely identified is the cedar, including in shipbuilding (the latter not being the case for cypress wood); also, cypress wood is less colourful than cedar, allowing rather clear distinction. Because of its location of origin, cedar could maybe be excluded: Morocco is generally regarded by Egyptologists as being too far away as a source for wood for ancient Egypt, but other coniferous species are very well possible.

4. Uses in Antiquity

The archaeological evidence presented suggests that cypress was used as an architectural material since the Predynastic period. Starting in the Old Kingdom, it was also used for making statues. Boxes and coffins are recorded since the Middle Kingdom. Two identifications seem particularly interesting: the small piece found in the abdominal mummy of Ramses II and a Nilometer base from the Byzantine Period. The first can be explained for its antiseptic properties and the second for its water resistance.

According to Lagriffe (1967: 67) cypress was used in the Kyphi composition. However, according to Baumann (1960: 88), who studied the Ebers Papyrus, which mentions the Kyphi composition, in this case cypress was confused with *Cyperus papyrus* L. (!). The medical use of cypress is at least confirmed for modern Egypt: around 1940, an infusion of cypress cones was employed in Egypt as an ailment for diabetes and rheumatism (Täckholm et al. 1941: 64).

The Greeks and Romans used cypress for shipbuilding (although Theophrastus does not mention it being used for this purpose[4]). While Alexander the Great, shortly before his death, was preparing a military expedition against Arabia, he made use of cypress (and ship owners) of Cilicia, Phoenicia, and Cyprus, as well as those that had been introduced to Babylon. And when Ptolemy wanted to build a fleet after the death of Alexander, he exercised strict control over the cypress and cedar forests of the mountains of Lebanon (Meiggs 1985: 134). The use of this species in the classical world seems to be well documented, for example, in the construction of monumental gates at the Parthenon, the temple of Epidaurus, and the basilicas of St. Peter and St. Sabina in Rome, and in its use (similar to that of cedar) by the Greeks for making chests, boxes, and coffins (Paraskevopoulou 1991: 165). The ashes of Athenians who had died during the first year of the Peloponnesian War were placed in cypress coffins (Meiggs 1985: 294), and this species has the reputation as having been used in the construction of the cross of Christ.[5]

References

Anglade E., 1988. *Musée du Louvre : catalogue des boiseries de la section Islamique*. Éditions de la réunion des musées nationaux, Paris.

Asensi Amorós M.V., 2002. Woods from ancient Egypt: The Ramesseum and the Valley of the Queens (18th Dynasty–Roman period): A preliminary report. In: Thiébault S. (ed.), *Charcoal Analysis: Methodological Approaches, Palaeoecological Results and Wood Uses*. [Proceedings of the second International Meeting of Anthracology, Paris, September 2000.] BAR International Series 1063. Archaeopress, Oxford, pp. 273-277.

Asensi Amorós M.V., 2003a. Annexe I : Analyse xylologique d'échantillons provenant des tombes du secteur du sanctuaire du Ramesseum. *Memnonia* 14, pp. 116-118.

Asensi Amorós M.V., 2003b. L'étude du bois et de son commerce en Egypte : lacunes des connaissances actuelles et perspectives pour l'analyse xylologique. In: Neumann K., Butler A. & Kahlheber S. (eds.), *Food, Fuel and Fields: Progress in African Archaeobotany*. [Proceedings of the 3rd International Workshop on African Archaeobotany.] Africa Praehistorica 15. Heinrich-Barth-Institut, Köln, pp. 177-186.

4 Theophrastus, 2003, V.7.1: 'The ship timbers are, in short, fir, pine and cedar'. In V.3.5, it is stated that the statues offered to the gods were made of juniper, cypress, jujube, boxwood, and olive roots for the smaller ones.

5 With four other species, including cedar (Baumann 1960: 89).

Bardinet T., 2008. *Relations économiques et pressions militaires en Méditerranée Orientale et en Libye au temps des pharaons : histoire des importations égyptiennes des résines et des conifères du Liban et de la Libye depuis la période archaïque jusqu'à l'époque ptolémaïque*. Etudes et mémoires d'égyptologie 7. Éditions Cybèle, Paris.

Barker H., Durleig R. & Meeks N., 1971. British Museum natural radiocarbon measurements VIII. *Radiocarbon* 13 (2), pp. 157-188.

Baumann B.B., 1960. The botanical aspects of ancient Egyptian embalming and burial. *Economic Botany* 14, pp. 84-104.

Boodle L.A., 1933. Unpublished report on file, Griffith Institute, Oxford. [Prepared for Howard Carter.]

Braidwood R.J., Jacobsen T., Parker R.A. & Weinberg S., 1951. Radiocarbon dates and their implications in the Near and Middle Eastern area. *Memoirs of the Society for American Archaeology* 8, pp. 1-65.

Brunton G. & Caton-Thompson G., 1928. *The Badarian civilisation and predynastic remains near Badari*. British School of Archaeology in Egypt, London.

CCA (Corpus Antiquitatum Aegyptiacarum) Martin, K., 1991. *Übersee-Museum Bremen*. Vol. 1. Die altägyptischen Denkmäler. Philipp von Zabern, Mainz am Rhein.

Callen G., 1976. *Les conifères cultivés en Europe*. Vol. 1. J.-B. Baillière, Paris.

Camus A., 1926. Le *Cupressus dupreziana*, cyprès nouveau du Tassili. *Bulletin de la société dendrologique de France* 58 (mai 1926), pp. 39-44.

Charpentier G., 1981. *Recueil de matériaux épigraphiques relatifs à la botanique de l'Egypte antique*. Éditions Trismégiste, Paris.

Collardet J. & Besset J., 1988. *Les bois commerciaux et leurs utilisations*. Vol. 1. Bois résineux (conifères). Éditions H. Vial, Dourdan.

Davies V.W., 1995. Ancient Egyptian timber imports, an analysis of wooden coffins in the British Museum. In: Davies V.W. & Shofield L. (eds.), *Egypt, the Aegean and the Levant, Interconnections in the Second Millenium B.C.* The Trustees of the British Museum, London, pp. 146-156.

Dietrich A., 2002. Les analyses de bois. In: Midant-Reynes B. & Buchez N. (eds.), *Adaïma I : economie et habitat*. Fouilles de l'Institut français d'archéologie orientale 45. Institut français d'archéologie orientale, Cairo, pp. 517-519.

Donadoni A.M., 1969. *Treasures of the Egyptian Museum*. Newsweek Books, New York.

El Hadidi M.N. & El Fayoumi, H.H., 1996. Catalogue of the archaeobotanical specimens in Cairo University herbarium. Vol. 1. Abu Sha'ar site, Red Sea coast, Egypt, season 1990. *Taeckholmia* 16, pp. 11-30.

Engelbach R., 1931. Ancient Egyptian woods. *Annales du Service des antiquités égyptiennes* 31, pp. 144.

Fakhry A., 1959. *The monuments of Sneferu at Dashur.* Vol. 1. The Bent Pyramid. General Organization for Government Printing Offices, Cairo.

Gale R., Gasson P., Hepper N. & Killen G., 2000. Wood. In: Nicholson P.T. & Shaw I. (eds.), *Ancient Egyptian Materials and Technology*. Cambridge University Press, Cambridge, pp. 334-371.

Garcia Esteban L., Guindeo Casasus A. & De Palacios P., 1996. *Maderas de coníferas: Anatomía de géneros*. Fundación Conde del Valle de Salazar, Madrid.

Germer R., 1988. *Katalog der altägyptischen Pflanzenreste der Berliner Museen.* Ägyptologische Abhandlungen 47. Otto Harrasowitz, Wiesbaden.

Hartung G., 2001. Organische Materialen: 5. 5. 1. Hölzer. In: Hartung U. (ed.), *Umm El-Qaab II: Importkeramik aus dem Friedhof U in Abydos (Umm el-Qaab) und die Beziehungen Ägyptens zu Vorderasien im 4. Jahrtausend v. Chr.* Archäologische Veröffentlichungen 92. Philipp von Zabern, Mainz am Rhein, pp. 313-317.

Hayes W.C., 1935. The tomb of Nefer-khewet and his family, the Metropolitan Museum of Art, The Egyptian Expedition 1934–1935. *Bulletin of the Metropolitan Museum of Art* 30 (2), pp. 17-36.

Hayes W.C., 1959. *The Scepter of Egypt*. Part 2. The Metropolitan Museum of Art, New York.

Killen G.P., 1980. *Ancient Egyptian Furniture*. Vol. 1. 4000–1300 BC. Aris & Phillips, Warminster.

Lagriffe L., 1967. *Le livre des épices, des condiments et des aromates.* Robert Morel éd., Paris.

Lansing A. & Hayes W.C., 1937. The Museum's excavations at Thebes: The Egyptian Expedition 1935–1936. *Bulletin of the Metropolitan Museum of Art* 32 (1), pp. 4-39.

Liphschitz N., 1998. Timber identification of wooden Egyptian objects in museum collections in Israel. *Journal of the Institute of Archaeology of Tel Aviv University* 25 (2), pp. 255-276.

Lucas A., 1936. The wood of the third dynasty ply-wood coffin from Saqqara. *Annales du Service des antiquités égyptiennes* 36, pp. 1-4.

Lucas A., 1989. *Ancient Egyptian Materials and Industries.* 4th ed., revised by Harris J.R. Histories & Mysteries of Man, London.

Luqma N., 2002. [Notes sur la méthode de fabrication de la statue en bois du Sheikh el-Balad.] In: Eldamaty M. & Trad M. (eds.), *Egyptian Museum Collections around the World.* Vol 1. Supreme Council of Antiquities, Cairo, pp. 84-89 [in Arabic].

Meiggs R., 1985. *Trees and Timber in the Mediterranean World.* Clarendon Press, Oxford.

Paraskevopoulou A.H., 1991. Variation of wood structure and properties of *Cupressus sempervirens* var. *horizontalis* in natural populations in Greece. *IAWA Bulletin* 12 (2), pp. 195-206.

Pliny the Elder, 1956. *Histoire naturelle.* Vol. 13. A. Ernout (ed., trans., annotator). Les belles lettres, Paris.

Plu A., 1985. Bois et graines. In: Balout L. & Roubet C. (eds.), *La momie de Ramsès II.* Éditions Recherche sur les civilisations, Paris, pp. 166-174.

Roehring C.H., 2002. *Life along the Nile: Three Egyptians of Ancient Thebes.* The Metropolitan Museum of Art, New York.

Sherif A.S. & el-Taife A., 1986. *Gymnosperms.* In: Jafri S.M.H. & el-Gadi A. (eds.), *Flora of Libya.* Al Faateh University, Faculty of Science, Tripoli.

Sloley R.W., 1953. Radiocarbon dating. *Journal of Egyptian Archaeology* 39, p. 115.

Strudwick N., 2006. *The British Museum Masterpieces of Ancient Egypt.* The British Museum Press, London.

Täckolm V., 1974. *Students' Flora of Egypt.* 2nd ed. Cairo University Press, Cairo.

Täckholm V., Täckholm G. & Drar M., 1941. *Flora of Egypt.* Vol. 1. Bulletin of the Faculty of Science 17. Fouad I University Press, Cairo.

Theophrastus, 2003. *Recherches sur les plantes.* Vol. 3. Books 5 and 6. Amigues S. (ed. & trans.). Les belles lettres, Paris.

Werker E., 1988. Wood. In: Rothenberg B. (ed.), *The Egyptian Mining Temple at Timna.* University College, London, pp. 232-235.

Wittmack L., 1910. Untersuchung altegyptischer Holzproben aus Abusir. *Sitzungsberichte der Gesellschaft Naturforschender Freunde zu Berlin* 4, pp. 181-192.

Zidan Y., Handoussa T., Hosni H. & El Hadidi N.M.N., 2006. The conservation of a wooden Graeco-Roman coffin box. *Preservation Science* 3, pp. 27-33.

Zohary M., 1973. *Geobotanical Foundations of the Middle East.* Geobotanica Selecta 3. Gustav Fischer Verlag, Stuttgart / Swets & Zeitlinger, Amsterdam.

Appendix 1

Overview of archaeological finds of *Cupressus* wood in Egypt.

Period	Date	Original location	Category	Item	Details	Current location
Prehistoric period						
		Badari	funerary object	several small pieces of worked wood	Identified as cypress or juniper (*Juniperus* sp.) (Brunton & Caton-Thompson 1928: 62 = Lucas 1989: 430 = Hartung 2001: 314 = Gale et al., 2000: 350).*	T4606; T3284
Predynastic Period (Naqada II–III)						
			architectural element	stake	(Dietrich 202: 517)	Adaïma, 1001/14.1
Old Kingdom						
	2033 cal BC	upper chamber, south pyramid of Snefru, Dahshur	architectural element	beam (?)	Made from *Cupressus* cf. *sempervirens;* a piece of wood removed for radiocarbon dating; although the report does not mention the object type (Barker et al. 1971: 160), it indicates that the samples were collected by Fakhry; in Fakhry's book (1959: 52), we find a mention of cedar beams without mention of analysis of wood (it could be a visual identification).	British Museum, London, BM 325
	4802 ± 210 cal BP	tomb of Snefru, south pyramid of Snefru, Dahshur	architectural element	beam (?)	A piece of wood removed for radiocarbon dating, referred to as 'cypress wood' by Sloley (1953: 115), citing the work of Braidwood et al. (1951: 53), without mentioning the object type, just the age.	
			statue	statue of Cheikh el-Beled	Statue made from *Cupressus sempervirens,* with the hand stick in tamarisk wood (Luqma 2002: 84-85). Before its scientific identification, this statue was referred to as sycamore fig wood by Liphschitz (1998: 267), following the identification by Donadoni (1969: 38); sycamore fig wood is, however, a totally different kind of wood, which cannot be mistaken for cypress.	Cairo Museum
	3rd Dynasty	Degrees pyramid, Saqqara	funerary object	coffin	Made of *Cupressus* sp. and of several elements (juniper, *Juniperus* sp.; jujube, *Ziziphus* sp.; cedar, *Cedrus* sp.; and Aleppo pine, *Pinus halepensis*), found in the Lauer excavations (Lucas 1936: 4 = Täckholm et al. 1941: 74 = Lucas 1989: 429, 440 = Gale et al. 2000: 352). Identifications made by Dr. L. Chalk of the Imperial Forestry Institute at Oxford, who states that one piece of the coffin (W.60) is probably *Cupressus* and another piece (W.752) must be *Juniperus* rather than a cypress.	
Middle Kingdom						
		Lahun	funerary object	coffin lid	Made from *Cupressus sempervirens* (Lucas 1989: 430 = Gale et al. 2000: 350).	

Period	Date	Original location	Category	Item	Details	Current location
		courtyard temple of Niouserrê, Abusir el-Meleq	funerary object	coffin of Sat-nefer	Made from *Cupressus sempervirens*: 'Koniferenholz, Abusir, bei der Pyramide des Neuserre, 5. Dyn. Grabung Borchardt 1903, det. Grosser: *Cupressus* spec. (Zipresse)'. Excavations Deutschen Orient Gesellschaft 1902, Bremen B 937, Übersee-Museum. Identification Grosser; CAA Martin 1991: 1). Germer (1988: 55) indicates a sample of cypress wood preserved in Berlin Museum (Ägyptisches Museum Berlin, n° S.Schweinfurth. 144). It may be a sample, not well dated, originating from the coffin of Sat-nefer.	Bremen B 937, Übersee-Museum
New Kingdom						
		tomb of Tutankhamun (Tomb 62), Valley of the Kings	funerary object	third coffin (two pieces)	Two fragments of the third coffin of Tutankhamun. Boodle (1933: 30) cites two pieces from 'Third outermost coffin'.	Royal Botanic Gardens, Kew, 253
	18th Dynasty, end of	Ramesseum, Tomb in the North Processional way	funerary object/ furniture?	fragmentary box	(Asensi Amorós 2002: 275)	J'"IV/A =APO.CN 01.To01
		tomb of Ramses II	funerary object	small, cube-shaped piece of wood	Abdominal filling of the mummy (Plu 1985: 167-168).	Cairo Museum
			funerary object	shabty (small statue figure of a funerary servant)	(Gale et al. 2000: 351)	Ashmolean Museum, Oxford, 1891.215a-b
		temple at Timna, Sinai, Wadi Arabah	worked wood	fragment of charcoal	(Werker 1988: 232)	
		tomb of Neferkhawt, Thebes, el-Asasif MMA 729	furniture	box	A small jewellery box of Rennofer (wife of Neferkhawt) with a lid in tamarisk and boxwood inlays (Hayes 1935: 29 = Lucas 1989: 429 = Gale et al. 2000: 337) (http://www.metmuseum.org/collections, last accessed in 2012); Hayes (1935: 29) states that the wood was identified by Professor S. J. Record of Yale University School of Forestry.	Metropolitan Museum of Art, New York, MMA 35.3.79
	18th Dynasty, early	tomb of Hatnefer and Ramose (below TT 71), Thebes, Sheikh Abd el-Qurna	funerary object	canopic box of Hatnofer	Hayes (1959: 226-227) says that the box is made of cypress, with a boxwood stick to close the two compartments that open into its upper part; the online database notes only wood (http://www.metmuseum.org/collections, last accessed in 2014).	Metropolitan Museum of Art, MMA 36.3.53
	18th Dynasty	tomb of Senenmut (TT71), Thebes, Sheikh Abd el-Qurna	weapons	bow	Found below entrance , deposit of hunting weapons, MMA 1935–1936; the online database notes the wood as 'conifer, cypress' (http://www.metmuseum.org/collections, last accessed in 2014).	Metropolitan Museum of Art, MMA 36.3.211
	18th Dynasty	tomb of Senenmut (TT71), Sheikh Abd el-Qurna	furniture	jewellery box	Small box said to be made from cypress wood, containing two compartments and decorated with boxwood inlays (Roehring 2002: 29) (http://www.metmuseum.org/collections, last accessed in 2014).	Metropolitan Museum of Art, MMA 36.3.199
	18th Dynasty, early	tomb of Hatnefer and Ramose (below TT 71), Thebes, Sheikh Abd el-Qurna	furniture	chair	Hatnefer chair, with boxwood and ebony inlays (Lansing & Hayes 1937: 13 = Roehring 2002: 32) (http://www.metmuseum.org/collections, last accessed in 2014).	Metropolitan Museum of Art, MMA 36.3.152

Appendix 1 continued

Period	Date	Original location	Category	Item	Details	Current location
Third Intermediate Period						
		Ramesseum, Tombs on the Sanctuary	funerary object	fragments of coffins	Several fragments of cedar coffins with fragments of cypress or juniper (Asensi Amorós 2003a: 117-118).*	Ramesseum, P.T 5/III; P.T 12; P.T 4/IV
Third Intermediate Period–Late Period						
	25th–26th Dynasty	pyramid of Menkaure, Giza	funerary object	coffin	The online database notes 'Wooden anthropoid coffin, restored from fragments, made to contain the body of King Menkaure (Mycerinus) of the Fourth Dynasty' (http://www.britishmuseum.org/research/collection_online, last accessed in 2014). The coffin is constructed from several different kinds of wood. The larger components are all of juniper and fir. Local timbers (sycamore fig and tamarisk) were used to make some of the dowels and tenons with which the planks are joined together. The coffin was not part of the original funerary equipment of Menkaure, but was made for a later restoration of his burial, as indicated by both stylistic and epigraphic evidence. However, Strudwick (2006: 264) indicates the use of cypress instead of fir and juniper for the coffin, and dowels made of *Ficus sycomorus* and *Tamarix*.	British Museum, BM 6647
Late Period						
	26th Dynasty	Deir el-Bahari	funerary object	coffin of Tabatha, daughter of Petamûn	Catalogued as sycamore fig wood, but subsequent analyses clearly showed that it was cypress (Engelbach 1931: 144 = Täckholm et al. 1941: 74 = Lucas 1989: 430).	Cairo Museum, 41059
Ptolemaic Period						
	4th cent. BC	Abusir el-Meleq	funerary object	coffin	*Cupressus* or *Juniperus* sp. (Wittmack 1910: 185 = Täckholm et al. 1941: 74-75 = Lucas 1989: 430 = Gale et al. 2000: 351)*	
			statue	cat statue	*Cupressus sempervirens* (Asensi Amorós, 2003: 185)	Museum of Mediterranean Archaeology, Marseille, AM 738
Graeco-Roman Period						
	1st cent. AD	funerary monument of Padykam, Hermopolis Magna (Tuna el-Gebel)	funerary object	coffin of Dhouty, father of Padykam, scribe of Khemenou	*Cupressus sempervirens* (Zidan et al. 2006: 28)	
		Saqqara	funerary object	coffin	*Cupressus sempervirens* (Asensi Amorós 2003: 186)	Museum of Mediterranean Archaeology, AM 264
		Thebes (?)	funerary object	mummy label	*Cupressus sempervirens* (Asensi Amorós 2003: 18)	Museum of Mediterranean Archaeology, AM 122
			funerary object	coffin	Cupressaceae; Killen (1980: 7) notes a coffin of the Graeco-Roman period collected in Egypt by Schweinfurth in 1884 and reported to be in the collections of Kew Museum (elements 1 and 4 in *Ficus sycomorus*, element 2 in *Abies*, element 3 in Cupressaceae, Davies 1995: 151); in my opinion, it could be the same specimen cited before from Abusir el-Meleq and dated to the Ptolemaic period.*	Royal Botanic Gardens, 30.1894/1.2.3.4

Period	Date	Original location	Category	Item	Details	Current location
Roman Period						
			funerary object	two mummy labels	(Davies 1995: 151)	British Museum; inventory numbers not cited
			architectural element	ornamental frame and plank in a construction	(Vermeeren, this volume)	
Roman or Byzantine Periods						
	300–700 AD	Abu Sha'ar (Red Sea)	worked wood	two pieces of worked wood	*Cupressus sempervirens* (El Hadidi & El Fayoumi 1996: 25, 28-29)	Cairo University Herbarium, UH-10-As.90 75.ii; UH As.90-10-166
	300–700 AD	Abu Sha'ar (Red Sea)	worked wood, furniture	several fragments of a box	*Cupressus sempervirens* (El Hadidi & El Fayoumi 1996: 22)	Cairo University Herbarium, UH As.90-10-25.i
Byzantine Period						
	end 5th cent.	Cairo	architectural element	nilometer base	*Cupressus sempervirens* is listed as having been used, along with pine, gum arabic (*Acacia nilotica* (L.) Willd. Ex Del.), the sycamore fig, and probably the juniper (Täckholm et al. 1941: 75 = Lucas 1989: 430-431).	
Islamic Period						
	11th cent.	Egypt	architectural element	decorative element of a molding framing	*Cupressus sempervirens* (Anglade 1988: 59; catalogue number 32)	Musée du Louvre, Paris, 4062
	mid-12th cent.	Egypt	architectural element	arch, fragment	*Cupressus sempervirens* (Anglade 1988: 74; catalogue number 45)	Musée du Louvre, 6348
		Cairo, Fostat, Istabl Antar, Excavations R. Gayraud (IFAO)	worked wood	three fragments of worked wood	work in progress for IFAO, identifications Asensi Amorós	Fostat, Istabl Antar, 891-1, 11001-1a 11001-1b
		Cairo, Fostat, Istabl Antar, Excavations R. Gayraud (IFAO)	architectural element	3 balusters	work in progress for IFAO, identifications Asensi Amorós	Fostat, Istabl Antar, 1005-4, 10500-4, 10201-6
		Cairo, Fostat, Istabl Antar, Excavations R. Gayraud (IFAO)	weaving or furniture element	needle or instrument (kohl stick?)	work in progress for IFAO, identifications Asensi Amorós	Fostat, Istabl Antar, 6660-2
		Cairo, Fostat, Istabl Antar, Excavations R. Gayraud (IFAO)	weaving element	spindle shaft	work in progress for IFAO, identifications Asensi Amorós	Fostat, Istabl Antar, 11516-1
		Cairo, Fostat, Istabl Antar, Excavations R. Gayraud (IFAO)	architectural element		work in progress for IFAO, identifications Asensi Amorós	Fostat, Istabl Antar, 1199-1
		Cairo, Fostat, Istabl Antar, Excavations R. Gayraud (IFAO)	writing material	writing board (?)	work in progress for IFAO, identifications Asensi Amorós	Fostat, Istabl Antar, 10966-1

* indicates doubtful occurrence

Reconstructing African agrarian prehistory by combining different sources of evidence: Examples for West African economic plants

Roger Blench

Kay Williamson Educational Foundation
8, Guest Road
Cambridge CB1 2AL, United Kingdom

E-mail: rogerblench@yahoo.co.uk

Abstract. The paper is an overview of the application of historical linguistics to the reconstruction of African agrarian history, and in particular the potential to develop hypotheses about species that have only a limited archaeobotanical record. It opens with a broad overview of the development of African crop repertoires and the methods of applying the tabulation of vernacular names to establishing the introduction and spread of particular species. It gives examples focusing on three species, the African locust bean, *Parkia biglobosa* (Jacq.) R.Br. ex G.Don, the aerial yam, *Dioscorea bulbifera* L., and pearl millet, *Pennisetum glaucum* (L.) R.Br. ssp. *glaucum*. The overall synthesis underlines the importance of evidence-based approaches, making available the data that support arguments concerning individual species.

Keywords: Africa, archaeobotany, linguistics, agriculture, economic plants

1. Introduction

Archaeobotany is sometimes treated as if it were an isolated discipline, disconnected from its broader place in the narrative of prehistory. But it is one element in the reconstruction of African agrarian and environmental history, which, more broadly conceived, is the linking of the present with the past. A credible narrative must account for the agricultural systems, crops, and useful plants in the current ethnographic record. This in turn has three elements:

Process	Explaining the processes that led to the adoption of agriculture and its change over the millennia
Chronology	Assigning credible dates to these processes
Species	Identifying the actual species involved in these processes

No one discipline can construct this narrative in isolation. Evidence from archaeobotany, linguistics, palaeoclimatology, genetics, and historical records needs to be combined to tell a convincing story. Since the year 2000 there has been a major expansion in archaeobotanical records for Africa, as well as a parallel growth in linguistic data. This growth has not been matched by a corresponding expansion of analytic results—in other words, narratives showing how individual data points can be linked together. The goal of this paper is to lay out some methodological principles for working toward these goals, and to take examples of some African economic trees, cereal crops, and tubers to explore how such an interdisciplinary approach would work in practice. It does not cover palaeoclimatology and genetics except in passing, but these remain crucial elements in an integrated story.

2. Background to African agrarian history

This section sketches the broad outlines of the history of plant use in Africa and in particular the chronological stratification of particular types of crops. Plants have presumably been exploited throughout human evolution, but a key step must have been the manipulation of 'wild' plants to make them more accessible or useful to humans. Ethnographic accounts show that yams, *Dioscorea* spp., were managed in the rainforest by foragers to improve yield (Bahuchet 1993; Dounias 1993) and that tree seeds were carried on migratory routes, leading to anthropic distributions, just as the baobab, *Adansonia digitata* L., marks the routes of Fulɓe pastoralists today, as they discard fruits after eating their pulp. Vines such as jumblebeads (*Abrus precatorius* L. 1753) (for decoration) and edible *Gnetum* spp. (Lowe 1984; Mialoundama 1993), associated with human settlement, occur virtually throughout the world, and they may well have spread following primary human movement out of Africa. Even if the analysis of starch grains and phytoliths becomes more advanced, distinguishing managed tubers from truly domestic species will remain problematic. West Africa is still home to a variety of marginally edible *Dioscorea* species, whose domestication status remains disputed (Hladik & Dounias 1993).

In the light of this, our image of African plant domestication has become focused on two broad areas, the Ethiopian Plateau and the West African region (both Sahel and forest). The Ethiopian Plateau was first identified by Vavilov (1992), while Murdock (1959) may well have been the first to draw attention to the importance of West Africa. Agriculture was previously thought of as very ancient in Africa, aided by some rogue radiocarbon dates, but more intensive archaeobotany treats agriculture as relatively recent (ca. 4500–3000 BP) with older dates being discounted (Kahlheber & Neumann 2007; Neumann 2003). This is of considerable significance for archaeolinguistics, because in most cases the major language families seem to be older than this (Blench 2006). As a consequence, it may well be that major domesticates *cannot* be properly reconstructed to proto-languages, which conflicts with many existing claims in the literature. However, if there has been a transfer of names for wild plants to the domestic types, as is certainly the case for yams, then a reconstruction cannot guarantee the antiquity of the farmed type (see discussion in Connell 1998).

Sub-Saharan African domesticates include sorghum, *Sorghum bicolor* (L.) Moench, pearl millet, *Pennisetum glaucum* (L.) Br. ssp. *glaucum,* finger millet, *Eleusine coracana* Gaertn., African rice, *Oryza glaberrima* Steud., t'ef, *Eragrostis tef* (Zucc.) Trotter, fonio, *Digitaria exilis* (Kippist) Stapf., iburu, *D. iburua* Stapf, cowpeas, *Vigna unguiculata* (L.) Walp., Bambara nuts, *V. subterranea* (L.) Verdc., Ethiopian pea, *Pisum abyssinicum* A. Br., guinea-yam, *Dioscorea rotundata* Poir., Hausa potato, *Solenostemon rotundifolius* Poir., okra, *Abelmoschus esculentus* (L.) Moench, oil-palm, *Elaeis guineensis* Jacq., and a host of minor species. A puzzle arises from the apparent early export of some crops to India; sorghum, pearl millet, finger millet and cowpeas apparently reach India by sea ca. 4000 BP (Blench 2003; Fuller 2003). In order to explain this, it has been proposed that Africa was an example of 'cultivation without domestication' (Haaland 1996), but equally well, we might simply have not yet found the earliest African examples of crops such as pearl millet and sorghum. Whatever the explanation, the transfer of the crops and the motivations and identity of the ship owners remains a puzzle. Other problematic transfers are early movements from the Sahel across the Sahara. Watermelon, *Citrullus lanatus* (Thunb.) Matsum. & Nakai, in the form with oily seeds, probably sesame, *Sesamum indicum* L., and tamarind, *Tamarindus indica* L., and perhaps the ben-oil tree, *Moringa oleifera* Lam., seem to have travelled this way at an early period, and pearl millet and sorghum.

Murdock (1959) pointed out the importance of vegecultural species from South-East Asia, notably the Musaceae, taro, *Colocasia esculenta* (L.) Schott, and the water-yam, *Dioscorea alata* L. Exactly how these reached Africa, and at what period, remains disputed. A single Musaceae phytolith from Cameroun dating to 2500 BP has been the source of much discussion (Mbida et al. 2000); if such crops arrived on the east coast earlier than this, then how did they cross a continent then devoid of cultivators occupying the appropriate area? Blench (2009) discusses the importance of the West African triploid plantains and other South-East Asian vegetative species and the puzzle of their somatic diversity. From roughly the sixth century onwards, a more conventional suite of Asian crops arrives on the East African coast including citrus, Asian rice, *Oryza sativa* L., probably sugar-cane, *Saccharum officinarum* L., cannabis, *Cannabis sativa* L. ssp. *sativa,* and betel, *Areca catechu* L. Few of these make their way far inland. Maghreb and Saharan domesticates arrive across the desert from around 2000 BP, including onions and jews' mallow, *Corchorus olitorius* L. The timing of the arrival of a separate suite of Near Eastern crops in Ethiopia is disputed. These crops include barley, *Hordeum vulgare* L., wheat, *Triticum* spp., lentils, *Lens culinaris* Medik., and fruits such as apricots, *Prunus armeniaca* L., and peaches, *P. persica* (L.) Batsch, but these stay in the Ethiopian highlands.

From the sixteenth century, new crops begin to arrive, most notably those from the New World, such as maize, *Zea mays* L., cassava, *Manihot esculenta* Crantz, pawpaw, *Carica papaya* L., guava, *Psidium guayava* L., and other species that have transformed African agriculture. The transfer was not only one way; okra, *Abelmoschus esculentus*, ackee, *Blighia sapinda* K.D.Koenig, yams, and oil-palm made their way across the Atlantic (Carney & Rosomoff 2009; Nunn & Qian 2010). In the twentieth century, modern agronomic species have displaced many traditional varieties, in a process that is still continuing. Vegetables such as tomatoes, *Solanum lycopersicum* L., carrots, *Daucus carota* L., and cabbage, *Brassica oleracea* L., at first not very successful, are now making important inroads into remoter rural areas. High-input species such as these also generally require chemical fertilizer, so traditional systems of manuring are also being discarded.

Only a very small proportion of these species are recorded in the archaeobotanical record, for reasons relating to preservation or to evolving techniques. Systematic flotation has changed the picture, but the analysis of starch grains is still only incipient. The more systematic application of techniques that have evolved in Pacific archaeology, notably the use of starch grains and phytoliths, would transform our image of African crop repertoires. The use of DNA has been applied patchily to some cereals but has yet to produce a major revelation.

3. Plant domestication and linguistic salience

Given the importance of West Africa as a centre for plant domestication and the broader role of Africa in human prehistory, surprisingly little attention has been paid to the linguistic evidence for plant use and domestication. Should we expect plant names to reconstruct? Blench (2007a) expresses a certain amount of scepticism about the possibilities of distinguishing loanwords from true reconstructions without much more reliable phonological data for each linguistic family or subgroup. Perhaps this is to be too demanding. A geographical cluster of cognate terms undoubtedly points to an interest in a particular plant; there will undoubtedly be lexical diffusion and semantic shift as well as genuine cognacy between related languages.

As our knowledge of African archaeobotany expands, it is clearly of interest to see if economic plants that are salient in the archaeological record can be matched against reconstructible linguistic roots. The key is understanding why and how plants are named. In most of sub-Saharan Africa the biotic environment is very rich, and any given ethnolinguistic group will name only a small proportion of the organisms they encounter, usually reflecting a combination of use and salience. Almost all larger mammals have specific names, but smaller species, particularly rodents, may be grouped together. Many insects are not distinguished, and usually only large, edible, or harmful fish are named. Similarly, plants acquire names when used. However, the great majority of names are not fundamental lexemes, but rather epithets, poetic descriptions similar to the English 'lily-of-the-valley'. These are generally of limited use in historical linguistics. For example, the dandelion (*Taraxacum* spp.) has a variety of related names in European languages. Set I refers to the belief that dandelions are a diuretic, Set II all etymologize as 'lion-tooth' from the appearance of the plant. Some of these can be parsed by speakers, while others are now-cryptic borrowings (Table 1).

Table 1. Related names for *Taraxacum* spp. in European languages.

Language	I	II
English	piss-a-bed	dandelion
Norwegian		løvetann
Danish	mælkebøtte	løvetand
German		Löwenzahn
French	pissenlit	dent de lion
Italian	piscialletto	dente di leone
Catalan	pixallits	dent de lleó
Spanish		diente de león
Portuguese		dente-de-leão
Welsh		dant y llew

Essentially Table 1 shows that in a restricted region, the names recognize the diuretic properties and the shape of the leaves and that the idea was borrowed between various branches of Indo-European at different times. Moving further east, these associations are lost, and dandelions are linked to deafness (Macedonia) or seen as the bringer of good news (Persian قاصدک, *qasedak*, which translates as 'small postman'). These are characteristic areal borrowings and calques and are of only limited historical use beyond folklore.

The first author to consider these issues for African crops was Portères (1958), but his lack of access to well-transcribed data in the 1950s made it problematic to reach any well-formed hypotheses. The comprehensive study by Blakney (1963) on names for banana was the first to try to link results with linguistic classification and archaeological data. Philippson & Bahuchet (1996) began the process of compiling and mapping Bantu names for major crop plants. Bostoen (2005, 2007a) has analysed the evidence for the reconstructed forms for economic trees, such as the oil-palm, *Elaeis guineensis,* in Bantu. Along similar lines, Blench (1996, 2003, 2006, 2007b,c, 2009) and Blench et al. (1997) have put forward a number of proposals for reconstructions of African economic

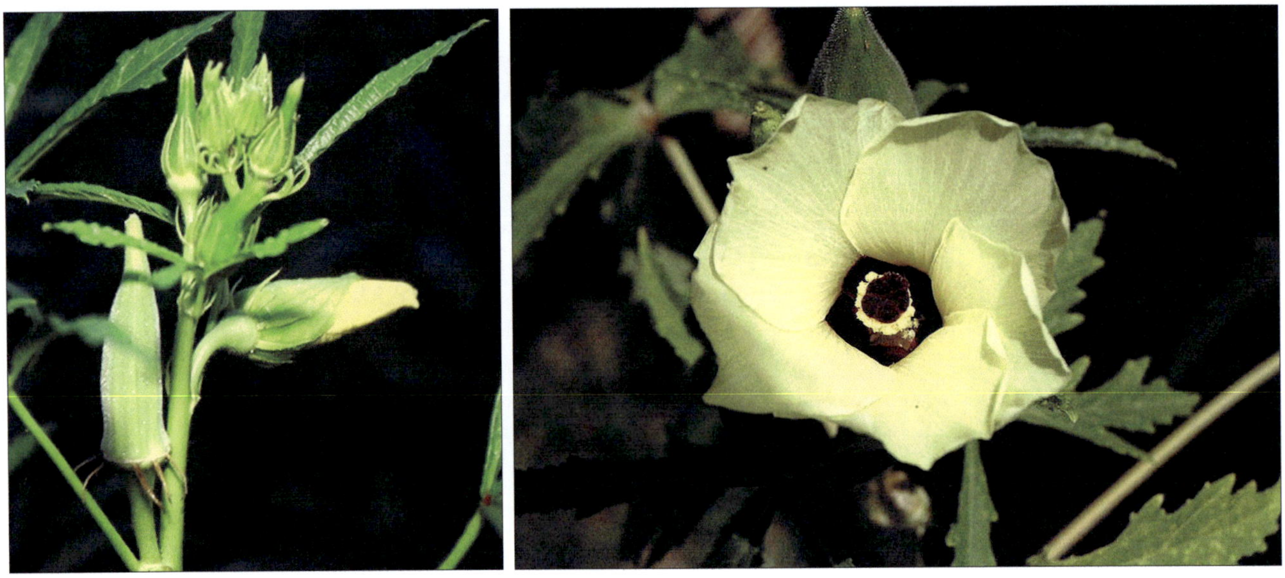

Figure 1. Okra, *Abelmoschus esculentus*.

Table 2. Names for okra, *Abelmoschus esculentus*, in the Niger Delta.

Lect	I	II	III
Defaka	ɔ́kʊrʊ		
Nkọrọ	ɔ́kuru		
Ịḅanị	ɔ́kʊrʊ́ʊ		
Kalaḅarị	ɔ́kʊrʊ		
Kịrịkẹ	ɔ́kʊru		
Nembe	ɔ́kʊrʊ́		
Akaha	ɔ́kurɔ		
Ḅụmọ		ɪkɪapʊ́	
Oporomọ		ɛkiyápú	
Oyakiri		ɪkɪyabʊ́	akɛnɛtá
East Tarakiri			akinãdá
East Olodiama		ɪkíyabʊ́	
Basan		ɪkíábʊ́	
Apọị		ɪkíábʊ́	
Iduwinĩ́	ɔ́kʊrʊ	ikɪabʊ	
Ogulagha		ɪkɪabʊ	
Gbaramatu		ɪkɪabɔ	
Fụrụpagha		ɪkɪabʊ	
Arogbo		ɪkɪabʊ	
Ogbe Ịjọ		ikɪabʊ	
Operemọ	ɔ́kʊrʊ		
Mein	ɔ́kurʊ		
Kunbo			akɪnɪtɛ́
Kabou			akɪnɪtí
West Tarakiri			akɛnɪtí
Ogboin		ekiapú	akinĩ́tá
Ikiḅiri			akinĩ́tá
Ekpetiama	ɔ́kʊrʊ		ɛkɛnɛtɛ́
Kolokuma	ɔkʊ́rʊ		ɛkɛnɛtɛ́
Gbaraịn	ɔ́kʊrʊ		
Oruma	ɔ́kʊrʊ		
Akịta	ɔ́kʊrʊ		
Biseni	ɔ́kʊrʊ dʊ́ʊ		

plants. Connell (1998) has explored the reconstructions for yams and oil-palms in a rather more limited area, the Cross River languages, and Maniacky (2005) has compiled the evidence for yam terms in Bantu languages.

The primary tool of palaeobiolinguistics (to use a felicitous term adopted by Cecil Brown) is the compilation of comparative names of plants and animals that may in principle have reconstructibility. This compilation should be undertaken across as many phyla and language families as possible, in order to ensure that loanwords are detected. It should then be possible to develop hypotheses as to which roots are related. Take the case of okra, an indigenous West African domesticate (Fig. 1). Table 2 shows the variety of names for 'okra' in the Niger Delta of Nigeria. Sometimes okra has two distinct names, but all of them fall into three related roots. This shows clearly that 'okra' cannot be reconstructed for proto-Ịjọ, but only for regional subgroupings, which is in line with the hypothesis that the Ịjọ reached the Niger Delta as a nomadic fishing people and only later adopted agriculture through contact with farmers and traders such as the Igbo (Williamson 1988).

Crops become salient when they are domesticated and thus categorically distinct from their wild relatives. In the case of trees the situation is more complex. With very few exceptions in recent times, trees are not truly domesticated, and they are not generally exotics. The discovery of their uses is a lengthy process, and is often stimulated by the evolution of technologies such as oil extraction. As a result, the name for a tree often spreads across a restricted zone within its broader natural distribution. For example, the shea butter tree, *Vitellaria paradoxa* C.F.Gaertn

(Fig. 2) is an important oil tree in savannah regions of West-Central Africa (Hall et al. 1996; see Fig. 3). Two subspecies are distinguished, subsp. *paradoxa*, which occurs from Senegambia to eastern RCA, and subsp. *nilotica*, which is distributed eastward into Uganda. Figure 3 shows its approximate distribution across the continent. Despite this, hard evidence for its importance and use in prehistory are slight. There is only one archaeobotanical record for shea, a fourteenth century testa from the medieval village of Saouga in Burkina Faso, while shea-butter production in the Sahel was recorded by Ibn Baṭṭuṭa in the fourteenth century (Neumann et al. 1998: 60).

Table 3 shows the linguistic evidence for a widespread Niger-Congo root for 'oil, fat' which, in some areas, has shifted to the specific meaning of the shea butter tree, *Vitellaria paradoxa*, and its oil. In western Niger-Congo languages this root seems to be a generic term for 'oil, fat'. However, with its occurrence in Kwa languages, it becomes specifically applied to the shea. In the southernmost Bantoid and A group Bantu areas, which are outside the ecological range of the shea, a savannah species, the word again has the more general meaning of 'oil, fat'. Figure 3 shows the distribution of this root superimposed on the 'natural' distribution of the shea. At its western and south-eastern distributions, this root simply means 'oil', but wherever the shea became a central oil-production species, the meaning 'shea' predominates.

The extraction of trees for timber or new products by Europeans caused the spread of common names from

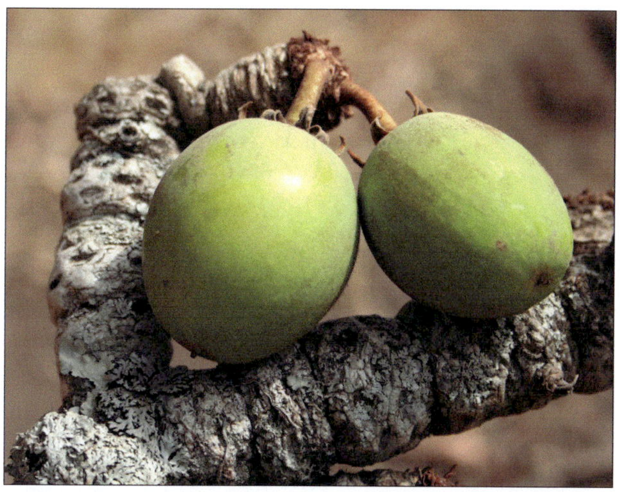

Figure 2. Fruit of the shea butter tree, *Vitellaria paradoxa*.

Table 3. Reflexes of *#ŋ-kpunu* 'oil'→ shea butter tree, *Vitellaria paradoxa*.

Family	Subgroup	Language	Attestation	Gloss	Source
Kru		Bete G	kpʉ́	*huile*	ALKrCI
Kru		Aizi	kpu	*huile*	ALKrCI
Atlantic	North	Mankanya	o-kərəʔ	oil	Wi07
Gur	Oti-Volta	Moba	kpàm	*graisse*	Ma75
Gbaya		Ɓodoe	kɔ̂l	shea	Ro08
Ubangian		Kpatiri	kpɔ	*graisse, huile*	Mo88
Kwa		Gonja	ŋ̀-kú	shea	Sn89
Kwa		Ga	ŋ̀kú	shea-butter	Kr99
Bijogo		Bijogo	ŋ-kiḓi	oil	Wi07
WBC	Igboid	Igbo	òkwùma	shea-butter	KW
WBC	Nupoid	Nupe	èkó	shea-butter nut	Ba14
EBC	Plateau	Obiro	òkʷô	shea tree	RMB
EBC	Plateau	Tinor	kɔ̏ŋɔ̏	shea tree	RMB
EBC	Plateau	Ake	kìkyɔ̏	shea tree	RMB
EBC	Plateau	Tarok	ìkíní	shea tree	RMB
Bantoid	Buru	Buru	ŋ̀ko	oil	RK
Bantoid	Momo	Ngwo	ŋ̄gúd	oil	Ey10
Bantu		CB	-gʊ́tà, -kʊ́tà	oil	G
Bantu	Jarawan	Doori	kólá	shea tree	MR

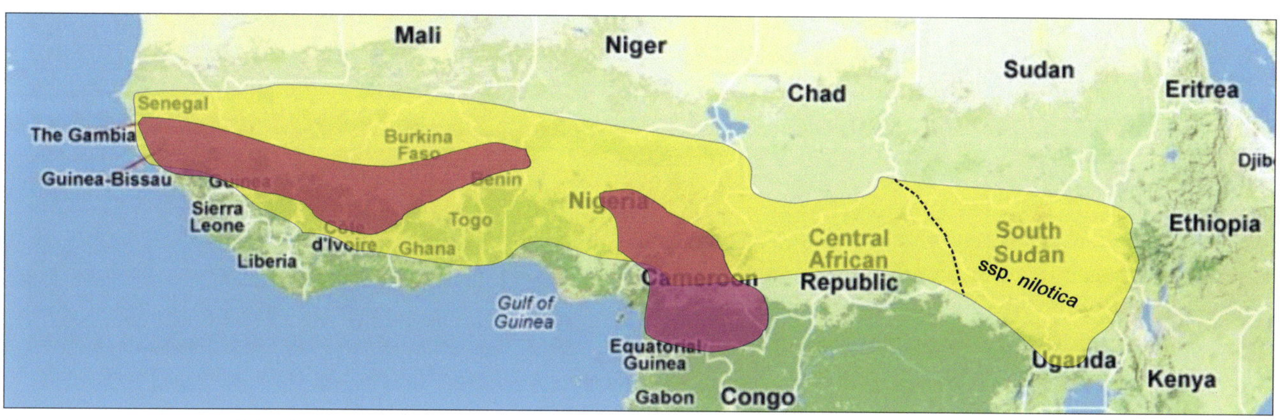

Figure 3. Distribution of the shea butter tree, *Vitellaria paradoxa*, and its vernacular names in Africa.

Table 4. Ghanaian names for the bush rubber tree, *Funtumia elastica.*

Branch/subgroup	Language	sg.	pl.
Tano	Twi	ɔ-fruntum	
	Nzema	ofuntum	
Gbe	Ewe	funtum	
Northern Guang	Gikyode	òfúntún	ìfúntún
Ka-Togo	Tuwuli	òfruntum	tùfruntum
Gur	Ntrubo	òfúrúntún	

Source: Irvine (1961)

the sixteenth century onwards. A product that came to be valued in the colonial era was rubber. Forestry officers were constantly on the lookout for species to compete with commercial rubber from the New World, *Hevea brasiliensis* Müll.Arg., and numerous vines and trees were tried out during this period. One group was the *Funtumia* spp. or bush-rubber trees, which came to have considerable commercial importance in Ghana (Burkill 1985: 151). Table 4 shows the names of the bush rubber tree, *Funtumia elastica* (Preuss) Stapf, in the Volta Region.

Again, such a uniform common name would not be expected across different Niger-Congo branches, and these terms probably only spread outward from the coast from the 1880s onward, when the rubber was first exploited.

With this in mind, it seems useful to present some examples of sets of related names for useful plants in West Africa and to compare these with whatever archaeobotanical evidence exists. Section 4 compiles a series of tables of related reflexes of what appear to be common roots, but makes no assumptions as to whether these are true phonological reconstructions or a mosaic of loanwords. In many cases, a mixture of the two is the most likely. Many established economic plants have so far not been recorded in any excavations, but their linguistic saliency provides hints for species to seek when sieving at a site. In some cases, there is also historical data to explore.

4. Case studies

4.1 Data presentation

Collating data from a large number of sources and presenting it in tables requires a considerable amount of compression to ensure the data is accurate and can be traced to the original. Each of the tables presents the phylum for the language, abbreviated as follows:

AA Afroasiatic
NC Niger-Congo
NS Nilo-Saharan

Two further columns present the family (Chadic, Kwa, etc.) and the subgroup (West, Oti-Volta). Two names call for comment: Volta-Niger (a proposed grouping of old Eastern Kwa, Yoruba, Nupe, etc. with the Gbe languages) and EBC, standing for Eastern Benue-Congo, the old Benue-Congo of Williamson (1971). The language name is the common name of a language. The attestation gives the original form cited in the source. The Gloss column is the definition as given in the source, given in the original language. This is done to avoid problems with overly convenient translations. The Source column gives the source in abbreviated form (see Appendix 1 for the expansions).

4.2 Species

4.2.1 African locust bean, *Parkia biglobosa*

The African locust bean, *Parkia biglobosa*, is currently one of the most important trees of the West African savannah (Hall et al. 1997). The seeds, flour, and pods are all eaten or used in construction (Fig. 4). Yet this tree barely features in the archaeobotanical record. A common root, something like *#-lona*, has developed in languages spread between Burkina Faso and Central Nigeria, which may point to an expansion of usage of products of the African locust bean tree after the major language families

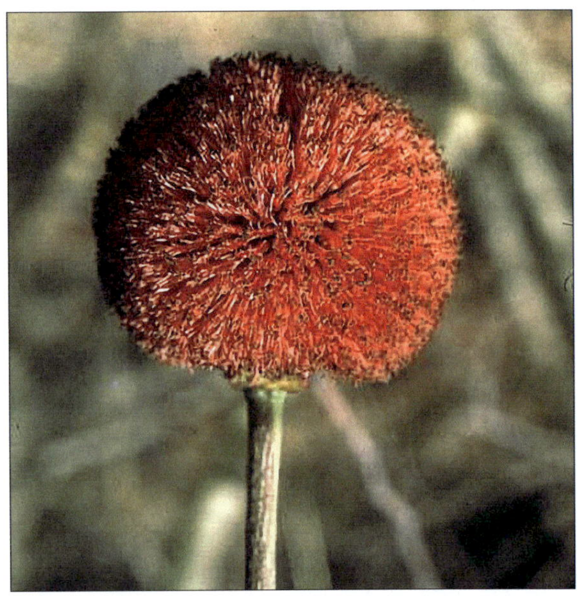

Figure 4. Flower of the African locust bean, *Parkia biglobosa*.

Table 5. The *#lona* root for the African locust bean, *Parkia biglobosa*.

Ph	Family	Subgroup	Language	Attestation	Gloss	Source
AA	Chadic	West	Hausa	ɗòòrowàà	locust tree	Ab49
AA	Chadic	Central	Bura	nônà	locust tree	RMB
NS	Saharan		Kanuri	runo	locust tree	Cy94
NC	Gur	Oti-Volta	Tamari	nuã	*néré*	Ma75
NC	Adamawa		Ɓəna [=Yungur]	rwoo	locust tree	RMB
NC	Volta-Niger	Yoruboid	Yoruba	iru	seed of ~	Ab58
NC	Volta-Niger	Nupoid	Nupe	elo	locust fruit	Ba14
NC	Volta-Niger	Nupoid	Gbagyi	olo	locust tree	RMB
NC	EBC	Kainji	Reshe	u-lo /tsu-	locust tree	RMB
NC	EBC	Kainji	Rin [Pongu]	ùrò	locust tree	RMB
NC	EBC	Kainji	Basa-Gumna	ulolo	locust tree	RMB
NC	EBC	Kainji	Cicipu	lɔ́ɔ pl. llɔ́ɔ	locust tree	SM
NC	EBC	Plateau	Iten	ɛ̀lool	locust tree	RMB
NC	EBC	Plateau	Cara	lɔl	locust tree	RMB
NC	EBC	Plateau	Ningye	urò	locust tree	RMB
NC	EBC	Plateau	Ashe	ì-rũ	locust tree	RMB
NC	EBC	Plateau	Idũ	ìrũ̀wầ	locust tree	RMB
NC	EBC	Plateau	Tinor	ì-rũ̀	locust tree	RMB
NC	EBC	Plateau	Hasha	ì-nɔn	locust tree	RMB
NC	EBC	Plateau	Ake	ìrɔ̃	locust tree	RMB
NC	Bantoid	Dakoid	Samba Daka	loom	locust tree	RMB
NC	Bantoid	Tivoid	Tiv	nune	locust tree	Ab40

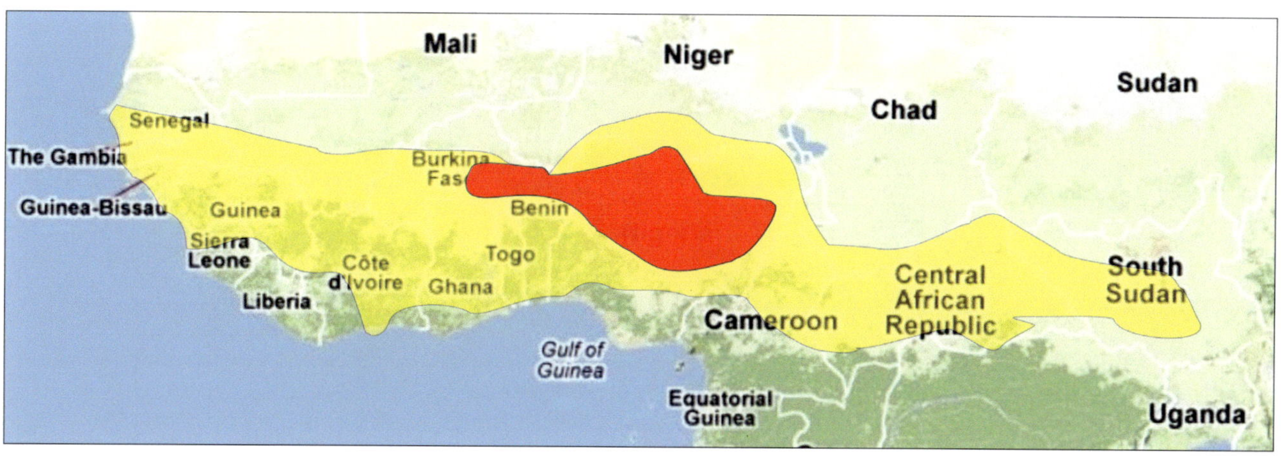

Figure 5. Distribution of the *#lona* root for the African locust bean, *Parkia biglobosa*.

were established, perhaps 2000 to 3000 years ago[1]. Table 5 shows the reflexes of a root for the African locust bean, *Parkia biglobosa*.

Figure 5 plots the geographical distribution of the *#lona* root for *Parkia biglobosa* (marked in red) against the natural distribution of the tree. The assumption is that the single widespread root, extensively borrowed, marks the discovery of techniques for processing of the seeds, thereby increasing its salience.

Reflexes are not widespread in either Chadic or Nilo-Saharan, and we can assume that the term originates with a Niger-Congo language. The density of terms in Central Nigeria, especially in Plateau, points to this region as the origin of the term. The original form seems to have had a V-prefix and a stem of the form *-lona/lono*, hence the nasalized forms. It would have then been borrowed both west into Gur and east into Adamawa languages. The Hausa form is isolated, and the *-waa* element

1 Lists of vernacular names are to be found in Burkill (1995) and Hall et al. (1997).

is assumed to be a suffix, with l→ɗ and the nasal in C_2 position becoming -r-.

4.2.2 Aerial yam, *Dioscorea bulbifera*

The aerial yam, *Dioscorea bulbifera*, known in Nigerian English as the 'up-yam', is cultivated for the bulbils that develop at the leaf axils (Fig. 6). In Africa, aerial yams are distributed from Senegambia to Kefa in south-west Ethiopia (Burkill 1985: 657 ff.; Martin 1974; Westphal 1975: 161). There are wild forms in both Africa and India, and Chevalier (1936) argues that the tuber was domesticated independently on both continents. The major morphological distinction between the quadrangular African forms and the ovoidal Indian types strongly suggest this. Chevalier claims that the Indian subspecies, *D. bulbifera* var. *birmanica*, was brought to the East African coast by the Arabs and to the West African coast by the Portuguese. There is no archaeological evidence for the aerial yam—and indeed it is unclear that it is possible in principle to distinguish domestic from wild types based on starch grains.

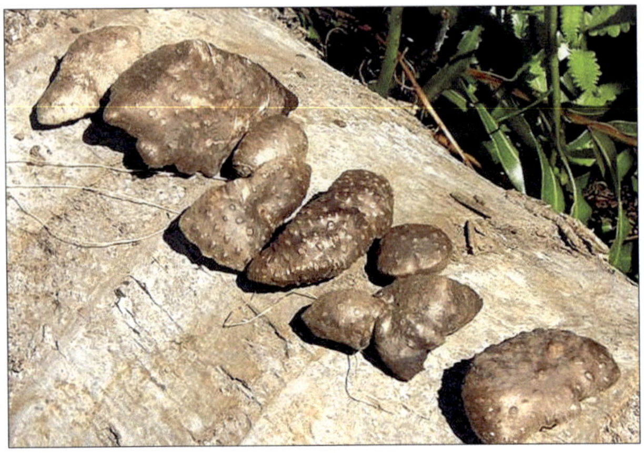

Figure 6. Aerial yam, *Dioscorea bulbifera*.

There is a widespread root, *#dom-*, applied to the indigenous aerial yam in Nigeria and adjacent regions of Cameroun, which was first identified in Williamson (1993). The species is not always well identified in the sources, so it may well be more widespread than this distribution suggests. The spread of this root seems to be coincident with Benue-Congo, while the attested forms in Volta-Niger languages tend to have a nasalized vowel.

Table 6. *#-dom-*, a root for aerial yam, *Dioscorea bulbifera*.

Family	Subgroup	Language	Attestation	Gloss	Source
Ijoid		P-Ijọ	*ɔtʊmʊ̃	aerial yam	KW
WBC	Edoid	Bini	udin	aerial yam	Ag86
Volta-Niger	Nupoid	Nupe	èdu	aerial yam	RMB
Volta-Niger	Igboid	P-Igboid	*-dʊ̃̌	aerial yam	KW
EBC	Kainji	tHun	rodiŋ tom	aerial yam	RMB
EBC	Kainji	ɛBoze	ri-don/a-	aerial yam	RMB
EBC	Plateau	Aten	ìtôm	aerial yam	RMB
EBC	Plateau	Berom	tòm	aerial yam	RMB
EBC	Plateau	Cara	i-tɔ	aerial yam	RMB
EBC	Plateau	Hyam	kpodom	aerial yam	RMB
EBC	Plateau	Izere	a-dom	aerial yam	RMB
EBC	Plateau	Idũ	ìdɛ̀m	aerial yam	RMB
EBC	Plateau	Ashe	ú-dũ̀	wild yam	RMB
EBC	Plateau	Nyankpa	ɛ̀dɔ̀m	aerial yam	RMB
EBC	Plateau	Hasha	ì-tum	aerial yam	RMB
EBC	Plateau	Sambe	ìntɔ́	aerial yam	RMB
EBC	Plateau	Horom	dùn	aerial yam	RMB
EBC	Plateau	Eggon	àdom	aerial yam	RMB
EBC	Plateau	Pe	atom	aerial yam	RMB
EBC	Lower Cross	Efik	édòmò	aerial yam	Co98
EBC	Central Delta	Abuan	ediom	aerial yam	Ga80
Bantoid	Mambiloid	Gembu	tūār	aerial yam	BC
Bantoid	Grassfields	Yamba	ntə́ntóŋ	k.o. small yam	RMB
Bantoid	Grassfields	Bafut	nìtu̅'ù	aerial yam	RMB
Bantoid	Grassfields	Ngomba	nɛtú'	aerial yam	RMB
Bantoid	Grassfields	Chufie'	tɔ́'ù	aerial yam	RMB
Bantu	C10	Aka	tombo	aerial yam	Thomas & Bahuchet (1991)

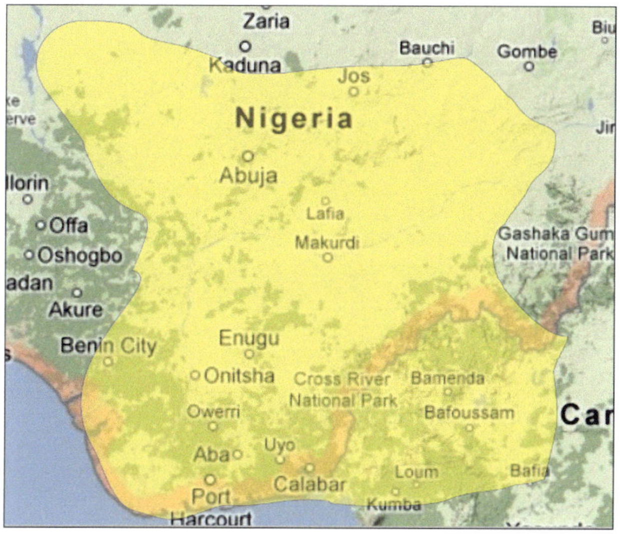

Figure 7. Distribution of the *#dom-* root for aerial yam, *Dioscorea bulbifera.*

The bilabial nasal marks the Ijoid form as a borrowing from Benue-Congo. Table 6 shows the attestations of the *#-dom-* root for the aerial yam.

It is striking that the western languages, predominantly Volta-Niger, have a high nasalized vowel in the stem, while the Benue-Congo languages usually have a mid-vowel plus a bilabial. This may be evidence that the plant was already known to speakers at the higher node before these subgroups split apart. Figure 7 shows the distribution of the *#dom-* root for *Dioscorea bulbifera.*

This suggests that the aerial yam first became salient for speakers of the language ancestral to Volta-Niger and Benue-Congo. If we assume that proto-Bantu can be dated to approximately 4000 BP, then proto-Benue-Congo must be several millennia older still. This suggests a date of at least 6000 BP.

4.2.3 Pearl millet, *Pennisetum glaucum* ssp. *glaucum*

Pearl millet is an important and ancient West African domesticate, and the first authors to claim an African origin were Koernicke & Werner (1885). Its wild relatives are found on the southern edge of the Sahara, and this area is usually considered to be its locale of domestication (Brunken et al. 1977; Chevalier 1932: 888-890). Archaeobotanical evidence for pearl millet in Africa is sparse but gradually improving (Nixon et al. 2011). Recent proposals situate this domestication event in the third millennium BC in the far western Sahel, perhaps in Mauritania and/or northeast Mali (D'Andrea et al. 2001; Fuller et al. 2007; Tostain 1998). Currently the earliest *Pennisetum glaucum* ssp. *glaucum* is that recorded at the Malian Neolithic sites of Karkarichinkat from 2500–2000 BC (Manning et al. 2010). Through much of subhumid West Africa, two significant subtypes are recognised, the tall, long-season semi-arid types (Hausa *maiwa*) and the short-season, humidity-tolerant types more characteristic of the Middle Belt (Hausa *gero*).

One of the emblematic sites of Nigeria, the Nok region, has produced a very large amount of pearl millet dated to 800–450 cal BC (or earlier?) (Kahlheber et al. 2009). Perhaps even more surprising is the fact that pearl millet was also cultivated in areas much farther south in regions that are now rainforest (Höhn et al. 2007). It is adopted early by the Berber and appears in Saharan oases by around the second century AD (after wheat) (Thanheiser et al., this volume) and makes some impression on southern Europe at the end of the Roman era and into the Middle Ages (Brunken et al. 1977). It also spreads from the Horn of Africa to India and Nepal some 4000 years ago.

Pearl millet is still grown as a ritual crop in the sandy, semi-arid areas of the eastern parts of coastal Ghana, though it has been completely displaced by maize as a staple. The polysemy of 'millet' and 'food' in a coastal language like Ga (in southeastern Ghana) is a striking indication of the former importance of millet in this region, since it has now almost entirely switched to growing maize as a staple.

Blench (under review) has tabulated the global evidence for vernacular terms applied to pearl millet. At least three major roots for pearl millet in Africa have been identified, two in the Bantu area, studied by Koen Bostoen (2007b), one in Central Nigeria (Longtau 2008), and one in the Central Chadic-speaking region in northern Cameroun (Richard Gravina, personal communication 2013). However, pearl millet appears also to have been transmitted to the Berber at an early period, since a single root appears all across Berber (Maarten Kossmann, personal communication 2013; Blench under review). However, in the proposed area of the western Sahel where domestication took place, there is a striking absence of common roots. This suggests that domestication may have been a slow and tentative process.

Table 7 shows a root for pearl millet, something like *#mar(d)-*, found across central Nigeria, which is widely attested in both Chadic and Plateau. The Chadic forms, such as Karekare *màrɗo*, look as if they contain the older form and that Hausa *maiwa* is probably a weakening of the second consonant. Zaar *màrwá* illustrates a transitional form in this weakening process. However, the Hausa name has subsequently been borrowed into other languages, such as Miya, which looks suspiciously similar. The Plateau names are all shorter and lack the *-rd-* sequence; hence it is most likely they were borrowed

Table 7. #mar(*d*)a, a root for pearl millet, *Pennisetum glaucum* spp. *glaucum*.

Ph	Family	Subgroup	Language	Attestation	Gloss	Source
AA	Chadic	West	Hausa	maiwa	millet	Ab49
AA	Chadic	West	Bole	mòrɗo	pearl millet	RS
AA	Chadic	West	Ngamo	mòrɗò	millet	RS
AA	Chadic	West	Geji	marɗa	millet	RMB
AA	Chadic	West	Ngizim	mar̄ɗû	millet	RS
AA	Chadic	West	Karekare	màrɗo	millet	RS
AA	Chadic	West	Kushi	mooɗo	millet	RMB
AA	Chadic	West	Miya	màywá	millet	RS
AA	Chadic	West	Mwaghavul	mààr	millet	RMB
AA	Chadic	West	Fyer	mar	millet	RMB
AA	Chadic	West	Sirzakwai	marɗay	millet	RMB
AA	Chadic	West	Zaar	màrwá	millet	BCa
NC	Adamawa	Yandang	Yoti	múri	millet	MR
NC	Kwa	Ga-Dangme	Ga	ŋmã̀ã̀	millet, food	Kr99
NC	Kwa		Adyukru	máy`	mil	ALKwCl
NC	Volta-Niger	Nupoid	Nupe	mã̀yì	millet	Ba14
NC	EBC	Plateau	Ninzo	amar	millet	RMB
NC	EBC	Plateau	Ningye	mwan	millet	RMB
NC	EBC	Plateau	Anib	àmɛ̂n	millet	RMB
NC	EBC	Plateau	Nyankpa	imala	millet	RMB
NC	EBC	Plateau	Ashe	i-ma	millet	RMB
NC	EBC	Plateau	Idũ	imara	millet	RMB
NC	EBC	Plateau	Shang	mara	millet	RMB
NC	EBC	Plateau	Jili	amo	millet	RMB
NC	EBC	Plateau	Sambe	tìk àmâr	millet	RMB
NC	EBC	Plateau	Kwaŋ	mɛr	millet	RMB
NC	EBC	Plateau	Yaŋkam	marak	millet	RMB
NC	EBC	Plateau	Tarok	imar	millet	Lo08
NC	EBC	Plateau	Sur	mər	millet	RMB
NC	EBC	Plateau	Pe	ime	millet	RMB
NC	Bantu	Jarawan	Mbula	mara	millet	MR
NC	Bantu	Jarawan	Mbat	máár	millet	MR

from Chadic at some point (which is also likely from an ecological point of view). It seems likely that the diffusion of this root records the increasing importance of pearl millet, which led to it being a ritual crop at the Nok sites. The finds of pearl millet at Nok in Central Nigeria are now dated to at least 2500 BP (Breunig 2014: 179 ff.), so we must assume it gained in importance some centuries prior to this date, given its significance for the Nok culture. Figure 8 illustrates the worldwide diffusion of pearl millet as it spread from the Sahel region of West Africa, to the Horn of Africa, on to South Asia and the fringes of China.

5. Conclusions

This paper has given some examples of the use of linguistics in developing hypotheses concerning the prehistory of African useful plants. Typically, vernacular names for domesticated plants in Africa *do not* reconstruct to deep-level proto-languages, only to more recent subgroupings, such as Bantu. Rather, they are regional or areal and tend to jump across language family and phylum boundaries. This is actually what we should expect, since the archaeobotanical data, where it exists, underlines a relatively late and sporadic, opportunistic agriculture in sub-Saharan Africa. Even trees and 'natural' vegetation tend to obey this rule; geographically based roots reflect not the initial domestication or introduction of a plant, but rather its transition into salience. This may reflect either a move toward becoming a dominant staple or the introduction of a new cultivar or technology (such as oil extraction). Patterns of lexical roots may also reflect palaeoclimate where they are discontinuous. Where forest disappears and then returns, names for trees may occur in fragmented patterns.

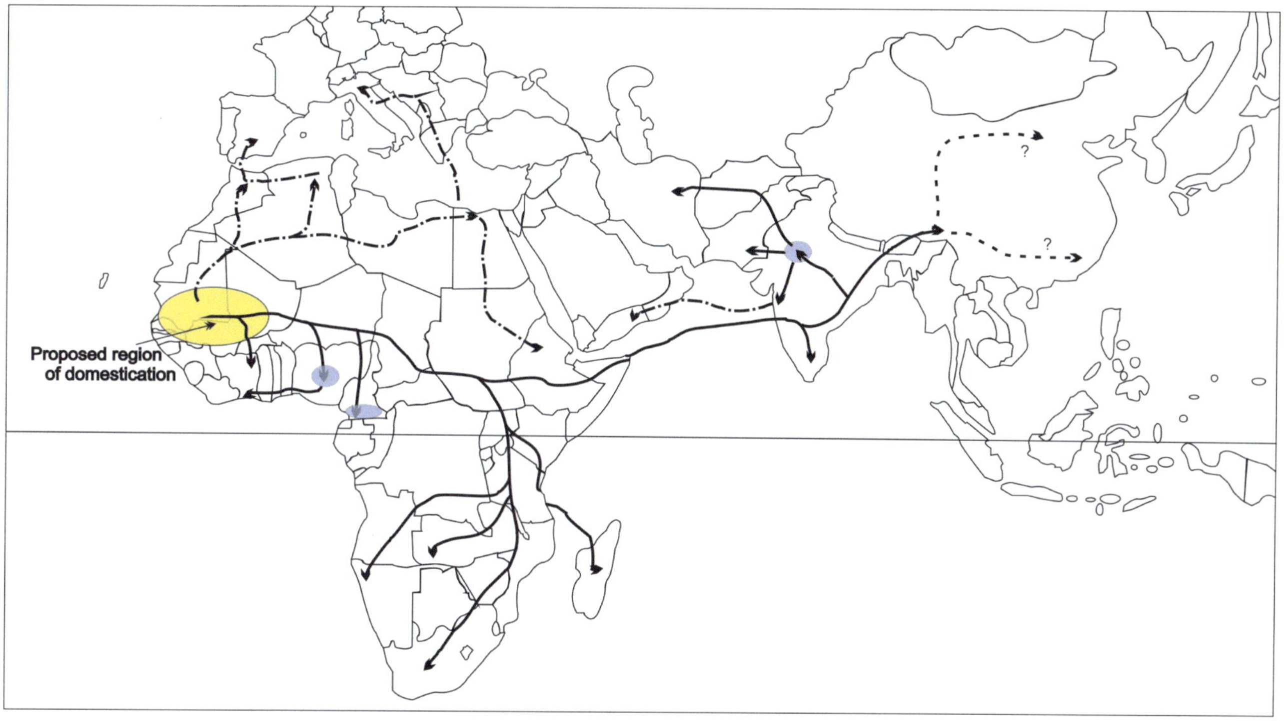

Figure 8. Worldwide diffusion of pearl millet, *Pennisetum glaucum* ssp. *glaucum*.

Linguistics has the potential to create hypotheses and to make suggestions for many species to fill the gaps in both origins and distribution, but it should be used in harmony with known archaeobotany. Hypotheses must be evidence-based—in other words, linguistic reconstructions should be supported by tables of evidence, not by assertions about starred forms. Archaeobotany in Africa is making slow progress, particularly in the area of vegetatively reproduced plants. Linguistics can contribute to the discussion by combining historical reconstruction with a sensible reading of the ethnographic data on agronomic practice.

References

Abraham R.C., 1949. *Dictionary of the Hausa Language*. University of London Press, London.

Abraham R.C., 1958. *Dictionary of Modern Yoruba*. University of London Press, London.

Agheyisi R.N., 1986. *An Ẹdo-English Dictionary*. Ethiope Publishing Corporation, Benin City.

Bahuchet S., 1990. The Aka pygmies: Hunting and gathering in the Lobaye Forest. In: *Food and Nutrition in the African Rain Forest*. Food Anthropology Unit 263. UNESCO, Paris, pp. 19-23.

Banfield A.W., 1914. *Dictionary of the Nupe Language*. The Niger Press, Shonga.

Blakney C.P., 1963. *On 'banana' and 'iron': Linguistic Footprints in African History*. Hartford Seminary Foundation studies in Linguistics, Hartford.

Blench R.M., 1996. Evidence for the inception of agriculture in the Nigeria-Cameroun borderland. In: Sutton J.E.G. (ed.). *The Growth of Farming Communities in Africa from the Equator Southwards*. Azania Special Volume 29-30. The British Institute in Eastern Africa, Nairobi, pp. 83-102.

Blench R.M., 2003. The movement of cultivated plants between Africa and India in prehistory. In: Neumann K., Butler A. & Kahlheber S. (eds.), *Food, Fuel and Fields: Progress in African Archaeobotany*. [Proceedings of the 3rd International Workshop on African Archaeobotany.] Africa Praehistorica 15. Heinrich-Barth-Institut, Köln, pp. 273-292.

Blench R.M., 2006. *Archaeology, Language and the African Past*. Altamira Press, Lanham.

Blench R.M., 2007a. Using linguistics to reconstruct African subsistence systems: Comparing crop names to trees and livestock. In: Denham T., Iriarte J. & Vrydaghs L. (eds.), *Rethinking Agriculture: Archaeological and Ethnoarchaeological Perspectives*. Left Coast Press, Walnut Creek, pp. 408-438.

Blench R.M., 2007b. *Enset* culture and its history in highland Ethiopia. In: Amha A., Mous M. & Savà G. (eds.), *Omotic and Cushitic language Studies: Papers from the IVth*

Cushitic-Omotic Conference, Leiden, April 10–12, 2003. Rüdiger Köppe Verlag, Köln, pp. 99-112.

Blench R.M., 2007c. The intertwined history of the silk-cotton and baobab in West Africa. In: Cappers R.T.J. (ed.), *Fields of Change: Progress in African Archaeobotany.* [Proceedings of the 4th International Workshop on African Archaeobotany.] Groningen Archaeological Studies 5. Barkhuis & Groningen University Library, Groningen, pp. 1-20.

Blench R.M., 2009. Bananas and plantains in Africa: Re-interpreting the linguistic evidence. In: Langhe E. De, Denham T. & Vrydaghs L. (eds.), Special issue: History of banana domestication. *Ethnobotany Research & Applications* 7, pp. 363-380.

Blench R.M. (under review). The contribution of vernacular names for pearl millet to its early history in Africa and Asia. Special issue: Millets. *Archaeological and Anthropological Science.*

Blench R.M., Williamson K. & Connell B., 1997. The diffusion of maize in Nigeria: A historical and linguistic investigation. *Sprache und Geschichte in Afrika* 14, pp. 19-46.

Bostoen K., 2005. A diachronic onomasiological approach to early Bantu oil palm vocabulary. *Studies in African Linguistics* 34 (2), pp. 143-188.

Bostoen K., 2007a. Bantu plant names as indicators of linguistic stratigraphy in the Western Province of Zambia. In: Payne D.L. & Peña J. (eds.), *Selected Proceedings of the 37th Annual Conference on African Linguistics*. Cascadilla Proceedings Project, Somerville, pp. 16-29.

Bostoen K., 2007b. Pearl millet in early Bantu speech communities in Central Africa: A reconsideration of the lexical evidence. *Afrika und Übersee* 89, pp. 183-213.

Breunig P. (ed.) 2014. *Nok: African Sculpture in Archaeological Context.* Africa Magna Verlag, Frankfurt.

Brunken J., Wet J.M.J. de & Harlan J.R., 1977. The morphology and domestication of pearl millet. *Economic Botany* 31, pp. 163-174.

Burkill H.M., 1985. *The Useful Plants of West Tropical Africa, Families A–D.* Royal Botanic Gardens, Kew.

Burkill H.M., 1995. *The Useful Plants of West Tropical Africa, Families J–L.* Royal Botanic Gardens, Kew.

Carney J. & Rosomoff R.N., 2009. *In the Shadow of Slavery: Africa's Botanical Legacy in the Atlantic World.* University of California Press, Berkeley.

Chevalier A., 1932. Les productions végétales du Sahara. *Revue de botanique appliquée et d'agriculture tropicale* 12, pp. 669-924.

Chevalier A., 1936. Contribution a l'étude de quelques espèces africaines du genre *Dioscorea*. *Bulletin du Muséum nationale d'histoire naturelle*, Series 2, 8 (6), pp. 520-551.

Connell B.A., 1998. Linguistic evidence for the development of yam and palm culture among the Delta Cross River peoples of southeastern Nigeria. In: Blench R.M. & Spriggs M. (eds.), *Archaeology and Language II.* Routledge, London, pp. 324-365.

Cyffer N. 1994. *English–Kanuri Dictionary.* Rüdiger Köppe, Köln.

D'Andrea A.C., Klee M. & Casey J., 2001. Archaeobotanical evidence for pearl millet (*Pennisetum glaucum*) in sub-Saharan West Africa. *Antiquity* 75, pp. 341-348.

Dounias E., 1993. The perception and use of wild yams by the Baka hunter-gatherers in South Cameroon rain forest. In: Hladik C.M., Hladik A., Linares O.F., Pagezy H., Semple A. & Hadley M. (eds.), *Tropical Forests, People and Food: Biocultural Interactions and Applications to Development.* Parthenon Publishing Group, Carnforth, pp. 621-623.

Eyoh J.A., 2010. Engwo Lexicon: Engwo–English, English–Engwo. Ngwo Language and Cultural Committee, Bamenda.

Fuller D.Q., 2003. African crops in prehistoric South Asia: A critical review. In: Neumann K., Butler A. & Kahlheber S. (eds.), *Food, Fuel and Fields: Progress in African Archaeobotany.* [Proceedings of the 3rd International Workshop on African Archaeobotany.] Africa Praehistorica 15. Heinrich-Barth-Institut, Köln, pp. 239-271.

Fuller D.Q., Macdonald K. & Vernet R., 2007. Early domesticated pearl millet in Dhar Nema (Mauritania): Evidence of crop processing waste as ceramic temper. In: Cappers R.T.J. (ed.), *Fields of Change: Progress in African Archaeobotany.* [Proceedings of the 4th International Workshop on African Archaeobotany.] Groningen Archaeological Studies 5. Barkhuis & Groningen University Library, Groningen, pp. 71-76.

Gardner I., 1980. *Abuan–English, English–Abuan dictionary.* Delta Series 1. University of Port Harcourt Press / Jos: Nigeria Bible Translation Trust, Port Harcourt.

Guthrie M. 1967–1971. *Comparative Bantu: An Introduction to Comparative Linguistics and Prehistory of the Bantu Languages.* 4 vols. Gregg International Publishers, Farnborough.

Haaland R. 1996. A socio-economic perspective on the transition from gathering to cultivation and domestication: A case study of sorghum in the middle Nile region. In: Pwiti G. & Soper R. (eds.), *Aspects of African Archaeology: Papers from the 10th Congress of the PanAfrican Association of Prehistory and Related Studies.* University of Zimbabwe Publications, Harare, pp. 391-400.

Hall J.B., Aebischer, D.P., Tomlinson H.F., Osei-Amaning E. & Hindle J.R., 1996. Vitellaria paradoxa*: A monograph.* School of Agricultural and Forest Sciences Publication 8. Bangor University, Bangor.

Hall J.B., Tomlinson H.J.F., Ogni P.I., Buchy M. & Aebischer D.P., 1997. Parkia biglobosa*: A monograph.* School of Agricultural and Forest Sciences Publication 9. Bangor University, Bangor.

Hérault G., 1983. *Atlas des langues Kwa de Côte d'Ivoire.* Vol. 2. ACCT and ILA, Abidjan.

Hladik A. & Dounias E. 1993. Wild yams of the African rain forest as potential food resources. In: Hladik C.M., Hladik A., Linares O.F., Pagezy H., Semple A. & Hadley M. (eds.), *Tropical Forests, People and Food: Biocultural Interactions and Applications to Development.* Parthenon Publishing Group, Carnforth, pp. 163-176.

Höhn A., Kahlheber S., Neumann K. & Schweizer A., 2007. Settling the rain forest—the environment of farming communities in southern Cameroon during the first millennium BC. In: Runge J. (ed.), *Dynamics of Forest Ecosystems in Central Africa during the Holocene: Past - Present - Future.* Palaeoecology of Africa: International Yearbook of Landscape Evolution and Palaeoenvironments 28. Taylor and Francis, London, pp. 29-41.

Irvine F.R., 1961. *Woody Plants of Ghana with Special Reference to Their Uses.* Oxford University Press, London.

Kahlheber S. & Neumann K., 2007. The development of plant cultivation in semi-arid West Africa. In: Denham T.P., Iriarte J. & Vrydaghs L. (eds.), *Rethinking Agriculture: Archaeological and Ethnoarchaeological Perspectives.* Left Coast Press, Walnut Creek, pp. 318-344.

Kahlheber S., Höhn A. & Rupp N., 2009. Archaeobotanical studies at Nok sites: An interim report. *Nyame Akuma* 71, pp. 2-17.

Koernicke F. & Werner H., 1885. *Handbuch des Getreidesbaues.* Verlag von Paul Paery, Berlin.

Kropp-Dakubu M.-E., 1999. *Ga–English dictionary with English–Ga Index.* Black Mask, Accra.

Longtau S.R., 2008. Linguistics and archaeology: Historical inferences from cognates for bulrush millet in Plateau and a review of research methodology. In: Ndimele O.-M., Udoh I.I.L. & Anyanwu O. (eds.), *Festschrift in Honour of Prof. Conrad Brann.* M & J Grand Orbit Communications Ltd. / Emhai Press, Port Harcourt, pp. 233-255.

Lowe J., 1984. *Gnetum* in West Africa. *Nigerian Field* 49 (1-4), pp. 99-104.

Manessy G., 1975. *Les langues Oti-Volta.* Société d'études linguistiques et Anthropologiques de France, Paris.

Maniacky J., 2005. Quelques thèmes pour «igname» en bantu. In: Bostoen K. & Maniacky J. (eds.), *Studies in African Comparative Linguistics with Special Focus on Bantu and Mande.* Musée royale de l'Afrique Centrale, Tervuren, pp. 165-189.

Manning K., Pelling R., Higham T., Schwenniger J.-L. & Fuller D.Q., 2010. 4500-year old domesticated pearl millet (*Pennisetum glaucum*) from the Tilemsi Valley, Mali: New insights into an alternative cereal domestication pathway. *Journal of Archaeological Science* 38 (2), pp. 312-322.

Marchese L. 1983. *Atlas linguistique Kru.* ACCT and ILA, Abidjan.

Martin F.W., 1974. *Tropical Yams and Their Potential.* Part 2. Dioscorea bulbifera. USDA Handbook 466. US Government Printing Office, Washington.

Mbida C.M., Van Neer W., Doutrelepont H. & Vrydaghs L., 2000. Evidence for banana cultivation and animal husbandry during the first millennium BC in the forest of southern Cameroon. *Journal of Archaeological Science* 27, pp. 151-162.

Mialoundama F., 1993. Nutritional and socio-economic value of *Gnetum* leaves in Central African forest. In: Hladik C.M., Hladik A., Linares O.F., Pagezy H., Semple A. & Hadley M. (eds.), *Tropical Forests, People and Food: Biocultural Interactions and Applications to Development.* Parthenon Publishing Group, Carnforth, pp. 177-182.

Moñino Y. (ed.), 1988. *Lexique comparatif des langues oubangiennes.* Geuthner, Paris.

Murdock G.P., 1959. *Africa: Its Peoples and Their Culture History.* McGraw Hill, New York.

Neumann K., 2003. The late emergence of agriculture in sub-Saharan Africa: Archaeobotanical evidence and ecological considerations. In: Neumann K., Butler A. & Kahlheber S. (eds.), *Food, Fuel and Fields: Progress in African Archaeobotany.* [Proceedings of the 3rd International Workshop on African Archaeobotany.] Africa Praehistorica 15. Heinrich-Barth-Institut, Köln, pp. 71-92.

Neumann K., Kalheber S. & Uebel D., 1998. Remains of woody plants from Saouga, a medieval West African village. *Vegetation History and Archaeobotany* 7, pp. 57-77.

Nixon S., Murray M.A. & Fuller D.Q., 2011. Plant use at an early Islamic merchant town in the West African Sahel: The archaeobotany of Essouk-Tadmakka (Mali). *Vegetation History and Archaeobotany* 20, pp. 223-239.

Nunn N. & Qian N., 2010. The Columbian exchange: A history of disease, food, and ideas. *Journal of Economic Perspectives* 24 (2), pp. 163-188.

Philippson G. & Bahuchet S., 1996. Cultivated crops and Bantu migrations in Central and Eastern Africa: A linguistic approach. *Azania* 29-30, pp. 103-120.

Portères R., 1958. Les appellations des céréales en Afrique : V : le Millet, Teff de l'Abyssinie. *Journal de l'agriculture tropicale et botanique appliqué* 5, pp. 454-463.

Roulon-Doko P., 2008. *Dictionnaire Gbaya–Français.* Karthala, Paris.

Snider K.L., 1989. *North Guang Comparative Wordlist.* Institute of African Studies, Legon.

Thomas, J.M. & Bahuchet S., 1991. *Encyclopédie des pygmées Aka : techniques, langage et société des chasseurs-cueilleurs de la forêt Centrafricaine (Sud-Centrafrique et Nord-Congo).* Vol. 1. Les Pygmées Aka. Fasc. 2. Le monde des Aka. Études Pygmées 4. Peeters, Louvain.

Tostain S., 1998. Le mil, une longue histoire: hypothèses sur sa domestication et ses migrations. In: Chastanet M. (ed.), *Plantes et paysages d'Afrique : une histoire à explorer*. Karthala, Paris, pp. 461-490.

Vavilov N.I., 1992. *Origin and Geography of Cultivated Plants*. Doris Löve (trans.). Cambridge University Press, Cambridge.

Westphal E.O., 1975. *Useful Plants of Ethiopia*. Centre for Agricultural Publishing and Documentation, Wageningen.

Williamson K. 1971. The Benue-Congo languages & Ịjọ. In: Sebeok T. (ed.). *Current Trends in Linguistics 7: Linguistics in Sub-Saharan Africa*. Mouton, Den Haag, pp. 245-306.

Williamson K. 1988. Linguistic evidence for the prehistory of the Niger Delta. In: Alagoa E.J., Anozie F.N. & Nzewunwa N. (eds.), *The Early History of the Niger Delta*. Helmut Buske, Hamburg, pp. 65-119.

Williamson K. 1993. Linguistic evidence for the use of some tree and tuber food plants in southern Nigeria. In: Shaw T., Sinclair P., Andah B. & Okpoko A. (eds.), *The Archaeology of Africa: Food, Metals and Towns*. One World Archaeology 20. Routledge, London and New York, pp. 104-116.

Wilson W.A.A., 2007. *Guinea Languages of the Atlantic Group: Description and Internal Classification*. Peter Lang, Frankfurt am Main.

Appendix 1

Acronyms and abbreviations used in the data tables, their source reference, and the language(s) they cover.

Acronym	Expansion	Language	Acronym	Expansion	Language(s)
Ab49	Abraham (1949)	Hausa	Kr99	Kropp-Dakubu (1999)	Ga
Ab58	Abraham (1958)	Yoruba	KW	Kay Williamson (personal communication 2004)	Ijoid
Ag86	Agheyisi (1986)	Ẹdo	Lo08	Longtau (2008)	Tarok
ALKCI	Hérault (1983)	Kwa	Ma75	Manessy (1975)	Oti-Volta
ALKrCI	Marchese (1983)	Kru	Mo88	Moñino (1988)	Ubangian
Ba14	Banfield (1914)	Nupe	MR	Mike Rueck (personal communication)	Nigerian languages
BC	Bruce Connell (personal communication)	Mambiloid	RMB	Author's fieldwork	West Africa
BCa	Bernard Caron (personal communication)	Zaar	Ro08	Roulon-Doko (2008)	Gbaya
Co98	Connell (1998)	Lower Cross	RS	Russell Schuh (personal communication 2008)	Chadic
Cy94	Cyffer (1994)	Kanuri	SM	Stuart McGill (personal communication 2013)	Cicipu
Ey10	Eyoh (2010)	Ngwo	Sn89	Snider (1989)	Guan
G	Guthrie (1967–1971)	Bantu	Wi07	Wilson (2007)	Atlantic
Ga80	Gardner (1980)	Abuan	Ga80	Gardner (1980)	Abuan

Loret V., 1892 [reprint 1975]. *La flore Pharaonique, d'après les documents hiéroglyphiques et les spécimens découverts dans les tombes*. Georg Olms Verlag, Hildesheim.

Murray M.A., 2000. Cereal production and processing. In: Nicholson P.T. & Shaw I. (eds.), *Ancient Egyptian Materials and Technology*. Cambridge University Press, Cambridge, pp. 505-536.

National Research Council. 1996. *Lost Crops of Africa*. Vol. 1: Grains. The National Academies Press, Washington.

Neef R., Cappers R.T.J. & Bekker R.M., 2012. *Digital Atlas of Economic Plants in Archaeology*. [Book and website of University Library Groningen.] Groningen Archaeological Studies 17. Barkhuis & Groningen University Library, Groningen.

Nesbitt M. & Samuel D., 1995. From staple crop to extinction? The archaeology and history of the hulled wheats. In: Padulosi S., Hammer K. & Heller J. (eds.), *Hulled Wheats*. [Proceedings of the first International Workshop on Hulled Wheats: Promoting the conservation and use of underutilized and neglected crops 4.] International Plant Genetic Resources Institute, Rome, pp. 41-100.

Peña L.Z., Peña-Chocarro L., Estévez J.J.I. & Urquijo J.E.G., 2003. Ethnoarchaeology in the Moroccan Jebala (Western Rif): Wood and dung as fuel. In: Neumann K., Butler A. & Kahlheber S. (eds.), *Food, Fuel and Fields: Progress in African Archaeobotany*. [Proceedings of the 3rd International Workshop on African Archaeobotany.] Africa Praehistorica 15. Heinrich-Barth-Institut, Köln, pp. 163-175.

Pettman A.J., Thanheiser U. & Churcher C.S., 2013. Provisions for the journey: Food production in the 'bakery' area of 'Ain el-Gazzareen, Dakhleh Oasis. In: Bagnall R.S., Davoli P. & Hope C.A. (eds.), *The Oasis Papers 6: Proceedings of the Sixth International Conference of the Dakhleh Oasis Project*. Dakhleh Oasis Project Monograph 15. Oxbow Books, Oxford, pp. 209-229.

Samuel D., 2000. Brewing and baking. In: Nicholson P.T. & Shaw I. (eds.), *Ancient Egyptian Materials and Technology*. Cambridge: Cambridge University Press, pp. 537-576.

Samuel D., 2009. Experimental grinding and ancient Egyptian flour production. In: Ikram S. & Dodson A. (eds.), *Beyond the Horizon: Studies in Egyptian Art, Archaeology and History in Honour of Barry J. Kemp*. Publications of the Supreme Council of Antiquities, Cairo, pp. 456-477.

Staubach S., 2005. *Clay: The History and Evolution of Humankind's Relationship with Earth's Most Primal Element*. The Berkley Publishing Group, New York.

Van der Veen M., 1996. The plant remains from Mons Claudianus, a Roman quarry settlement in the Eastern Desert of Egypt—an interim report. *Vegetation History and Archaeobotany* 5, pp. 137-141.

Van der Veen M., 2011. *Consumption, Trade and Innovation: Exploring the Botanical Remains from the Roman and Islamic Ports at Quseir al-Qadim, Egypt*. Journal of African Archaeology Monograph Series 6. Africa Magna Verlag, Frankfurt.

Vartavan C. de, Arakelyan A. & Amorós V.A., 2010. *Codex of Ancient Egyptian Plant Remains/Codex des restes végétaux de l'Egypte ancienne*. 2nd ed. Academic Books, London.

Vincentini O., Maialetti F., Gazza L., Silano M., Dessi M., Vincenzi M. de & Pogna N.E., 2007. Environmental factors of celiac disease: Cytotoxicity of hulled wheat species *Triticum monococcum, T. turgidum* ssp. *dicoccum* and *T. aestivum* ssp. *spelta*. *Journal of Gastroenterology and Hepatology* 22, pp. 1816-1822.

Wetterstrom W., 1995. Foraging and farming in Egypt: The transition from hunting and gathering to horticulture in the Nile Valley. In: Shaw T., Sinclair P., Andah B. & Okpoko A. (eds.), *The Archaeology of Africa: Food, Metals and Towns*. One World Archaeology 20. Routledge, London and New York, pp. 165-226.

Woenig, F., 1886 [reprint 1971]. *Die Pflanzen im alten Aegypten: Ihre Heimat, Geschichte, Kultur und ihre mannigfache Verwendung im sozialen Leben in Kultus, Sitten, Gebräuchen, Medizin und Kunst*. Philo Press, Amsterdam.

An investigation of taphonomic processes by means of digital image analysis of *Phoenix dactylifera* L. seeds from Roman Karanis (Fayum, Egypt)

Federica Fantone

Faculty of Archaeology
Leiden University
P.O. Box 9515
2300 RA Leiden, the Netherlands

E-mail: f.fantone@arch.leidenuniv.nl

Abstract. Taphonomic processes such as disposal patterns and contamination have important effects on the representativeness of the archaeobotanical record. The interpretation of these processes may be complicated by the fact that indigenous species are consistently represented in the dataset. Measurements of subfossil seeds of the date palm (*Phoenix dactylifera*) have mostly been used to identify different cultivars. The question posed here is: Can these measurements also be used to infer taphonomic processes? In an attempt to answer this question, digital image analysis was performed on photographs of 2730 subfossil date palm seeds from Roman Karanis, located on the edge of the Fayum Oasis in Egypt. For this purpose, 30 parameters that describe the size and shape of each specimen were computed. The surface versus length distribution provided the best discrimination. Although measurements of date palm seeds from different archaeological contexts show a wide range, they appear to correlate with (a) animal activity and (b) clear evidence of contamination (New World species) associated in the same context. It proved difficult to isolate single depositions. The results may be used to evaluate the potential of different archaeological contexts for investigating processes of mixing and reduction by taking into account both biotic and abiotic factors.

Keywords: taphonomy, archaeobotanical archive, representativeness, bias, digital images

1. Introduction

1.1 Representativeness and bias of the archaeobotanical archive

The representativeness of the archaeobotanical archive depends on the factors and processes that introduce variation and bias in the botanical composition and on the accuracy of their reconstruction. The reconstruction of how these factors act, which is the subject of taphonomy (Gifford 1981; Schiffer 1972, 1976, 1983), can be described through the processes of dispersal/deposition, mixing/contamination, and reduction (Cappers & Neef 2012: 173).

Preliminary studies that attempted to distinguish the effects of these processes on subfossil plant remains showed several potential sources of bias, including (a) the characteristics of plants and plant parts, their use, and their disposal; (b) the action of biotic and abiotic factors; (c) the preservation conditions; and (d) the sampling, processing, and analysis procedures (Miksicek 1987). Fundamental contributions stressed the potential of different types of archaeological contexts in the identification of activity patterns (Dennell 1972, 1974, 1976).

During the past decades work has focused on the reconstruction of possible routes of entry of plant remains into the archaeobotanical record. Only recently has emphasis been placed on the analysis of desiccated and mixed assemblages (Cappers 2006; Newton 2004; Rowley-Conwy 1994; Smith 2003; Thanheiser 2008; Van der Veen 2007, 2011). Special attention has been drawn to the characterization of food processing and consumption waste, fodder, fuel, mud bricks, and dung layers, and several contributions have reported marks of animal activity on seeds and fruits (among others, Cappers & Hamdy 2007; Panagiotakopulu & Buckland 2010; Van der Veen 2011).

Experiments on modern plant assemblages have provided a point of reference for the identification of rodent gnawing marks as incisor-sized impressions on the surface of seeds of, for example, watermelon (*Citrullus lanatus* (Thunb.) Matsum. & Nakai), bottlegourd (*Lagenaria siceraria* (Molina) Standl.), and juniper (*Juniperus*) (Cox & Van der Veen 2008; Gasser & Adams 1981: 185). Small holes present on the endocarp of subfossil sugar date (*Balanites aegyptiaca* (L.) Del.) and on the fruit of Nile acacia (*Acacia nilotica* (L.) Willd. ex Del.) have been related to insect activity (Cappers 2006: 56-63). However, while the attention being paid to such evidence is increasing, its impact on the representativeness of the archaeobotanical archive is not always sufficiently discussed. For instance, only recently has a clear point been made concerning the identification of small fragments of the central part of date palm seeds that have been partly eaten by rodents (Cappers & Neef 2012: 198; Neef et al. 2012: 55). These leftovers are very fragile and fragmented and may easily escape identification, with possible implications for the representativeness of this species. Finds of date palm seeds and fruits are almost unknown in the Predynastic period (before 3000 BC) and very rare until the Middle Kingdom (c. 2040–1640 BC); most originate from the New Kingdom (c. 1550–1070 BC) and post-Pharaonic periods (after 332 BC). But even in the New Kingdom, the proportion of finds is small in comparison with the importance of the fruit at that time as suggested by iconographic and linguistic evidence (Murray 2000). Given the high decay resistance of date palm seeds, these low proportions may relate to the sampling method, the type of context selected for study—for example, compare the difference in ubiquity and concentration of remains found at the Stone Village with those from other areas in Amarna (Workmen's Village, Grid 12, and the House of Ranefer) (Clapham & Stevens 2012; Stevens & Clapham 2010)—and the difficulty in identifying fragmented remains. It is clear that in this case establishing the contribution of the species to the past food economy becomes problematic.

The analysis of the food economy can be tackled by adopting a qualitative as well as a quantitative approach. A qualitative or semi-quantitative approach can be based on the list of the identified taxa. These data can be evaluated in terms of presence/absence (as well as proportions), bearing in mind that both morphology and taphonomy can play a role in their interpretation. A fully quantitative approach can be based on counts of plant parts and on the measurements of morphometric features of specimens.

In archaeobotanical research, morphometric measurements have been applied to, for example, the study of cereal grain kernels in order to pinpoint potential morphological changes due to domestication, charring, and environmental conditions (Willcox 2004). Measurements of date palm seeds have been used to distinguish between wild and domesticated species and to identify different cultivars through the comparison of subfossil material with known recent cultivars (Beech 2003; Beech & Shepherd 2001; Clapham & Stevens 2009; Nesbitt 1993; Terral et al. 2012).

To what extent can the measurements of morphometric parameters be applied to the investigation of taphonomic processes? A working hypothesis is that the analysis of digital images of seeds or fruits can reveal patterns that can be related to taphonomic processes in connection with (a) the overall botanical composition of the samples and (b) the stratigraphy and characterization of the related contexts.

The aim of this chapter is to contribute to the investigation of the processes of deposition, reduction, and contamination in different types of archaeological contexts by means of digital image analysis of date palm seeds. After a short overview on taphonomic factors and processes, a case study on subfossil seeds retrieved from the north-eastern portion of the Roman settlement of Karanis (Fayum, Egypt) is presented in order to examine the potential of different contexts (granary, street, house) in identifying taphonomic processes.

1.2 Taphonomy

Taphonomic factors are biotic and abiotic in nature. Their activities form the processes of dispersal/deposition, mixing/contamination, reduction, and preservation (Cappers & Neef 2012: 173). Dispersal depends on the characteristics of plants, such as seed production and dispersal strategies; the features of diaspores (weight, size, shape, decay resistance); crop processing procedures; and plant economic value and use (Hillman 1981).

As far as cultivated species are concerned, a first distinction may be made between crops that require several further processing stages after harvesting, such as cereals, and crops that do not. Among this latter group are several fruit plants. Once parts of these fruit plants enter the settlement, their intra-site dispersal is influenced not only by plant seed production, but also by patterns of food processing, consumption, and waste disposal.

For example, some fruits (or false fruits) produce several hundreds of seeds, such as the fig (700–2200 seeds/false fruit) or the pomegranate (250–400 seeds/fruit), although a considerable variation in the actual number produced per (false) fruit may occur (Cappers et al. 2007: 136). Others produce only one seed, which is also very decay-resistant, such as the date palm. Some fruits are eaten in their entirety, including the inner part, while of others the inedible part is discarded after the edible part has been consumed. This is the case with large one-seeded fruits, which can thus be dispersed across the site and/or end up in domestic refuse being (re)deposited within the settlement in unused or not visible parts of buildings (Rowley-Conwy 1994: 30) or outside in small or large middens.

The archaeological context is fundamental to the representativeness of the archaeobotanical record. Depending on the way deposition occurs, the archaeological context may be primary or non-primary. Only in the case of primary deposits can the botanical composition be considered to be representative of the function of the context itself (Hubbard & Clapham 1992). Furthermore, as a basin of deposition, the context itself affects the probability that a deposit will be subject to particular processes, while its dimensions, in combination with the rate of deposition, may influence the concentration of items (Schiffer 1983).

Mixing mostly takes place at the moment of deposition, for example, in connection with waste disposal activities. Sweepings of ash layers, debris of architectural material accumulated inside rooms, dung of animals kept inside rooms along with their fodder and bedding, and residue of food preparation and consumption may all become incorporated into the domestic waste, forming assemblages with a different degree of mixing.

The mixing of subfossil material with recent material is regarded as contamination, a definition that may also be applied to other phenomena, for example, if one or more minor components reduce the degree of homogeneity of otherwise pure samples (Jones & Halstead 1995: 103).

The probability of plant material being preserved in different archaeological contexts depends on the state of preservation and the activity of reducing factors (termed preservation environments, after Miksicek [1987]). In dry environments such as Egypt, plant material is preserved in two main modes: desiccation and charring. Mineralization can also occur in the presence of deposits with a high level of phosphate (Green 1979). The incorporation of charred material into the archaeobotanical archive depends on whether or not that plant material has come in contact with and is resistant to fire (Hillman 1981: 139-140; Van der Veen 2007: 978-979). Possible routes of entry are (a) intentional or casual use of plant remains as fuel, which includes that of inedible parts of fruits; (b) intentional or unintentional destruction of stored food and fodder by fire; (c) accidental burning during food preparation; (d) cleaning up of storage facilities or destruction of diseased crops by means of deliberately set fires; and (e) burning of rubbish.

Wind deflation, roaming animals, and insects are responsible for the reduction of the archaeobotanical archive. Wind deflation acts selectively. It produces a surface of heavy particles, while the smallest particles are usually blown away until new depositions take place and cover them (Cappers et al. 2007: 136). Browsing animals that are allowed to visit trash deposits may cause a considerable reduction in volume, while burrowing animals, such as small rodents, may intensively consume plant remains and thus bias the archaeobotanical record. Some plants are eaten in their entirety (e.g. pulse seeds). Others are gnawed only in part, to gain access to the softer inner part (e.g. safflower [*Carthamus tinctorius* L.] and olive [*Olea europea* L.]). Often only short fragments of the gnawed specimens are left; these then risk being overlooked or interpreted as being the result of a different process.

Sometimes patterns in animal activity can explain the composition of the archaeobotanical record. Experimental studies have shown that rodents gnaw and consume large amounts of food soon after it is discarded and until a new supply becomes available nearby (Gasser & Adams 1981: 107; Nowak & Paradiso 1983: 746). In addition, rodents may carry food back to their burrows, where it can be consumed as well as stored and/or discarded. Consequently, places selected as burrows may turn out to be rich in plant remains, but, inconveniently for the archaeobotanist, their composition is filtered by the rodents' food preferences (Miller 1996: 53).

All this does not mean that highly mixed contexts should be avoided; they should, instead, be understood in their components because they can tremendously enlarge the plant spectrum.

1.3 Why date palm seeds?

The analysis of date palm seeds presents some advantages for the investigation of taphonomic processes. Their ready availability (both in the past and today) and their high decay resistance favour the abundant presence of date palm seeds in the archaeological record. The ways in which the inedible part of the fruits is discarded and accumulated in domestic waste affects the distribution in the archaeological context and, thus, may reflect patterns of consumption and waste disposal. The marks left on the seeds that are not completely eaten by roaming and burrowing animals (especially rodents) frequently allow us to identify their activity. The high visibility of the specimens allows for a large number to be recovered—even if hand-picking is (systematically) performed in the field by non-specialists—and to be identified. The ease of quantifying whole items and thereby their contribution to the reconstruction of the past food economy mostly stems from our ability to relate each seed to only one fruit.

Date palm seeds also present some disadvantages connected with the difficulty of distinguishing between subfossil and recent seeds, in particular within assemblages that do not present apparent evidence of contamination (which in Egypt usually means New World taxa) and that contain date palm seeds in similar preservation condition. For this reason, one criterion relating to the state of preservation of the seed coat that is used to identify contamination—namely, the exterior colour—may be not sufficient: the outside part of the date palm seed can also turn blackish after short exposure to sunshine. Furthermore, short seed fragments represent a quantification problem because it is difficult to convert them to complete items and thus infer their contribution to the food economy.

1.4 The site

Karanis (currently known as Kôm Aushim) lies about 80 km south-west of Cairo, on the east side of the main Cairo–Fayum highway, at the northern edge of the Fayum Oasis in Egypt. The site is located on a limestone ridge at the rim of the desert between the wadi to the north and the cultivated area to the south and covers 60 ha. The settlement was occupied from the mid-third century BC to the fifth to sixth centuries AD or later (Gazda 1983: 1).

The first systematic excavations were carried out by Michigan University in the 1930s (Boak & Peterson 1931). The settlement seems to have reached its apex of occupation in the Early Roman period, with an estimated total population of about 4000 inhabitants during the first half of the second century AD (Rathbone 1990: 132, on the basis of Rylands Papyrus IV 594). After this stage, it underwent a decline in the late second century, followed by phases of alternating reconstruction and additional decline (Husselman 1979: 9). Later reconstructions mostly occurred on abandoned and decayed structures, although in some cases buildings were continuously occupied and structural alterations and re-occupations took place over the course of time before their final abandonment. The inhabitants' habit of dumping rubbish in the streets and in abandoned rooms within buildings caused the occupation level to rise to such an extent that often ground floors turned into basements and new floors were added

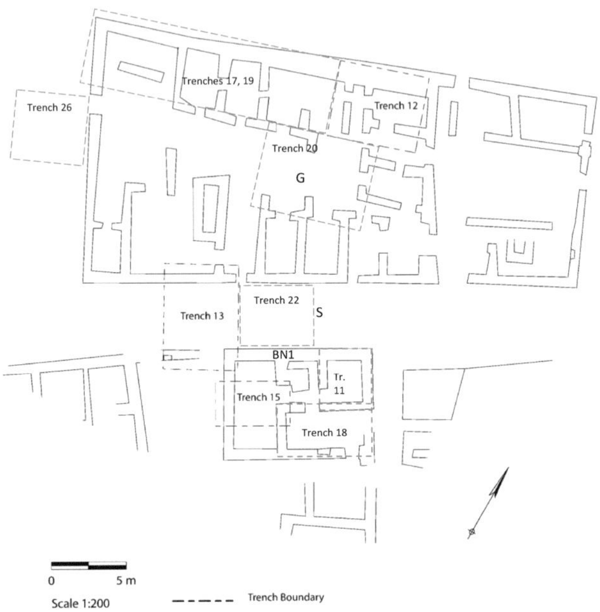

Figure 1. North-eastern Karanis. The map shows the investigated area with some of the trenches of the 2009 and 2010 excavation seasons mentioned in the text. G: Granary; S: Street; BN1: Building N1. (Map courtesy of the Fayum Project.)

above those no longer being used (Husselman 1979: plans 22-43).

The case study concerns the north-eastern extent of the settlement, where investigations have taken place since 2009 as part of the Fayum Project, jointly conducted by the University of California, Los Angeles; Rijksuniversiteit Groningen; and University of Auckland (Wendrich 2010) (Fig. 1). In this area the main ancient street runs south-west to north-east, in the direction of the plateau where the Late Roman–Early Coptic cemetery is located. Since 2009, excavations have focused on a sector of the main street and some extant buildings. Among these, a structure with a central courtyard surrounded by storage units (bins) has been interpreted as a granary (Trenches 12, 17, 19, 20, and 26). The storage units are subdivided into three to five subunits (Trench 12) by means of plastered mud brick walls, while underground units present beneath the hallway are vaulted with mud bricks and made accessible through hatches. Bin floor and walls were covered with several layers of plaster, which sometimes showed traces of repair.

On the opposite side of the street (Trenches 13 and 22), in an area already in use, a building (N1) was constructed of limestone blocks. The prior use of the area is suggested by a thin ash layer that extended under the north-western wall and which was levelled to serve as a surface at the moment of construction. Initially intended to function as a walled courtyard, the building underwent architectural alterations and may have served a residential purpose before a possible temporary abandonment. After the construction of the internal walls, the ground floor was divided into three rooms. The western (Trench 15), north-eastern (Trench 11), and south-eastern (Trench 18) rooms were connected by a corridor, while a staircase gave access to an upper level. The building may have lost its function before the definitive abandonment. Post-abandonment robbery activities are suggested by three trenches dug around the base of the staircase.

Botanical samples were retrieved from the granary, the house, and the street through a combination of judgement sampling and systematic on-site dry sieving. Judgement sampling favoured selected contexts (mixed deposits in the midden as well as storage facilities, ash layers, and mud brick architectural material), while systematic on-site sieving produced evidence from each stratigraphic unit.

Systematic on-site sieving may be biased toward the representation of taxa whose remains are in general well visible, such as complete date palm seeds. In addition, the application of this sampling method has the advantage of providing a potentially high representativeness of the sampled population.

The variety of the archaeological contexts and the availability of material from each of the investigated *loci* offer the opportunity to attempt an investigation of taphonomic processes through the analysis of morphometric parameters of date palm seeds.

2. Material and methods

A total of 441 samples was taken from the *loci* under investigation in the granary (Trenches 12, 17, 19, and 20), the street (Trenches 13 and 22), and the house (Trenches 11, 15, and 18) during the 2009 and 2010 excavation seasons. The soil from each stratigraphic unit was dry-sieved in the field through 2.0 mm mesh sieves by trench supervisors and fieldschool students (field-sieved samples). Sometimes more than one sample was taken from within the same unit in accordance with the distribution of the botanical remains in the deposit. In addition, 25 soil samples from selected contexts were dry-sieved in the lab on ≥ 2.0 and 0.5 mm mesh. The remains contained in the ≥ 2.0 mm sieve residue were fully processed by the author; a subsample of a standard unit of 400 ml was taken from the 0.5 mm sieve residue. The analysis of the soil samples allowed the characterization of the deposits (Fantone 2012), but their detailed description is beyond the scope of this paper.

For the digital images, 2730 seeds from the field-sieved samples were selected according to the following criteria: (a) completeness and (b) diversity. In order to obtain reliable measurements, only complete specimens of each sample were digitally photographed. At least two types were noted: thin and elongated, and short and slim (Fig. 2).

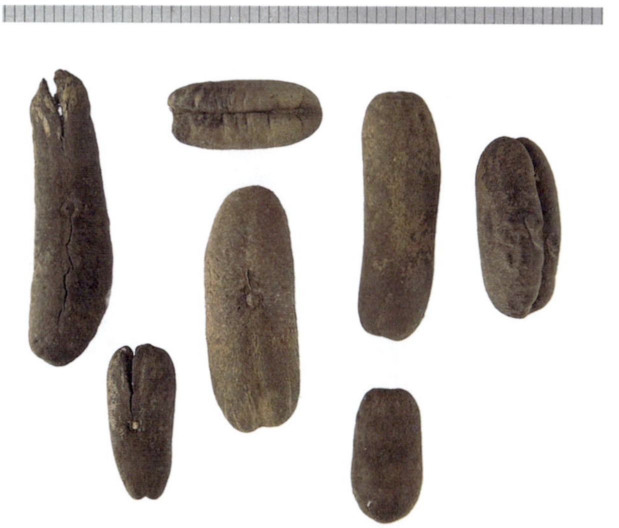

Figure 2. Date palm seeds belonging to two different varieties: (a) thin and elongated; (b) short and slim (reproduced with permission from Neef et al. 2012: 53).

Specimens with clear rodent gnawing marks were isolated and photographed separately. Three main categories of rodent gnawing were distinguished: seeds only superficially damaged by incisor-sized impressions; half-eaten seeds; gnawed seeds of which only the central part was left (Fig. 3).

All photos were made on a Zeiss stereo microscope at 3× confirm lens and 6× magnification using a Zeiss 5 megapixel digital camera. The seeds were positioned with the furrow uppermost. A transparent background and an artificial light were used in order to avoid creating a shadow. Because contact between two adjacent seeds had to be avoided so that they would not be measured as one

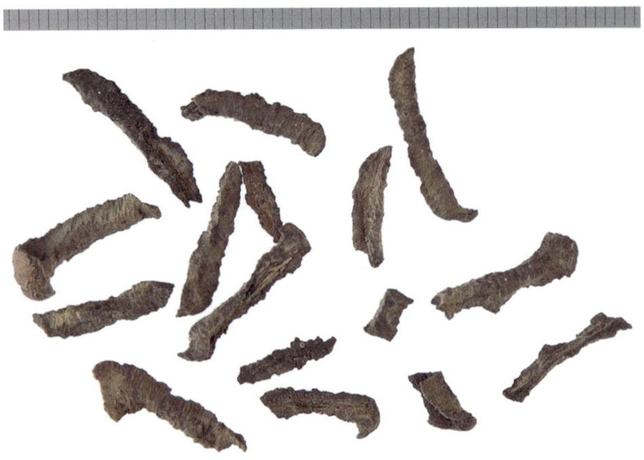

Figure 3. Date palm seed gnawed by small rodents, of which only the central part remains (Trench 12, U. 88) (reproduced with permission from Neef et al. 2012: 54-55).

single object, a maximum of between four and six seeds were photographed together.

Zeiss KS400 3.0 software was used to digitally measure the parameters from the photos. Calibration was performed with a length of 13–15 mm. To enable the program to measure the whole visible (ventral) surface of each seed and to accommodate the artificial light present, an exposure time of 770 ms was selected. With a standard gain on 3, the scan resolution of the final photograph was set to 3900 × 3090 pixels.

Out of the 30 parameters recorded by the software, only surface, width, and length were used for the current analysis. Data were scatter-plotted as length versus width as well as surface versus length. The best separation was found by plotting the surface against the length. In the surface/length plot, each specimen was attributed to the context of origin (trench number). The analysis was conducted by identifying groupings of specimens that might be related to processes of reduction, contamination, and mixing in connection with (a) the characteristic features of the specimens that might be indicative of one or more of such processes (e.g. conditions of preservation, gnawing marks, discolouration); (b) the overall botanical composition of the related samples, which includes evidence of mixing with recent material; and (c) the stratigraphy of the related archaeological context. One particular context (the granary in Trench 12) was selected for a more detailed analysis.

3. Results and discussion

The distribution of the three measurements was found to be quite uneven in all trenches, and some outliers could be observed. Three main groups were distinguished and associated with the processes of, respectively, reduction, contamination, and mixing (Fig. 4).

At bottom left, group A, representing reduction, includes the smallest charred seeds (surface between 6 and 40 mm^2) and some desiccated gnawed specimens, of which mostly only the central part remains, ranging from 10 to 17 mm in length. Associated with those are, for all trenches, some charred specimens that turned out to be shorter than the gnawed ones but of comparable calculated surface. These findings agree with what is known about the effects of charring on the dimensions of date palm seeds, which tend to decrease in length and increase in width (Rowley-Conwy 1987). Most of the charred seeds were from the street (Trench 13) and the house (Trench 18).

In the central part of the plot, the larger group, B, connected with contamination, includes the majority of the seeds from the south-eastern room of building N1 (Trench 18), which range in surface from 40 to 122 mm^2. The whitish colour of the seed coat and its excellent state of preservation suggest a recent date (Neef et al. 2012: 52). Arguments based on botanical as well as stratigraphic grounds seem to support this hypothesis. The botanical composition basically produced evidence of dung and architectural decay and contained only sparse botanical remains, in line with the archaeological interpretation of the deposits as (re)depositions of trash during post-abandonment activities and levellings (no occupational layer was reached). With few exceptions, the date palm seeds included in this group were found in association with clear evidence of contamination. These included New World crop plants such as peanut (*Arachis hypogea* L.), sunflower (*Helianthus annuus* L.), and maize (*Zea mays* L.), as well as some more problematic plants that are indigenous or have been introduced to Egypt in antiquity. Among these are sorghum (*Sorghum bicolor* (L.) Moench) and rice (*Oryza sativa* L.). In addition, skins of onion (*Allium cepa* L.) were found at the north-west corner of Trench 18 on the sealed surface of a trash deposit. Their colour and twisted shape also appears to be indicative of a recent date. No comparable signs of contamination were found in the granary bin (Trench 12).

At top right, group C, representing mixing, includes the biggest and longest specimens (surface from 135 to 165 mm^2), which are more widely scattered and were found in different contexts (granary, street, western and south-eastern rooms of building N1). These specimens

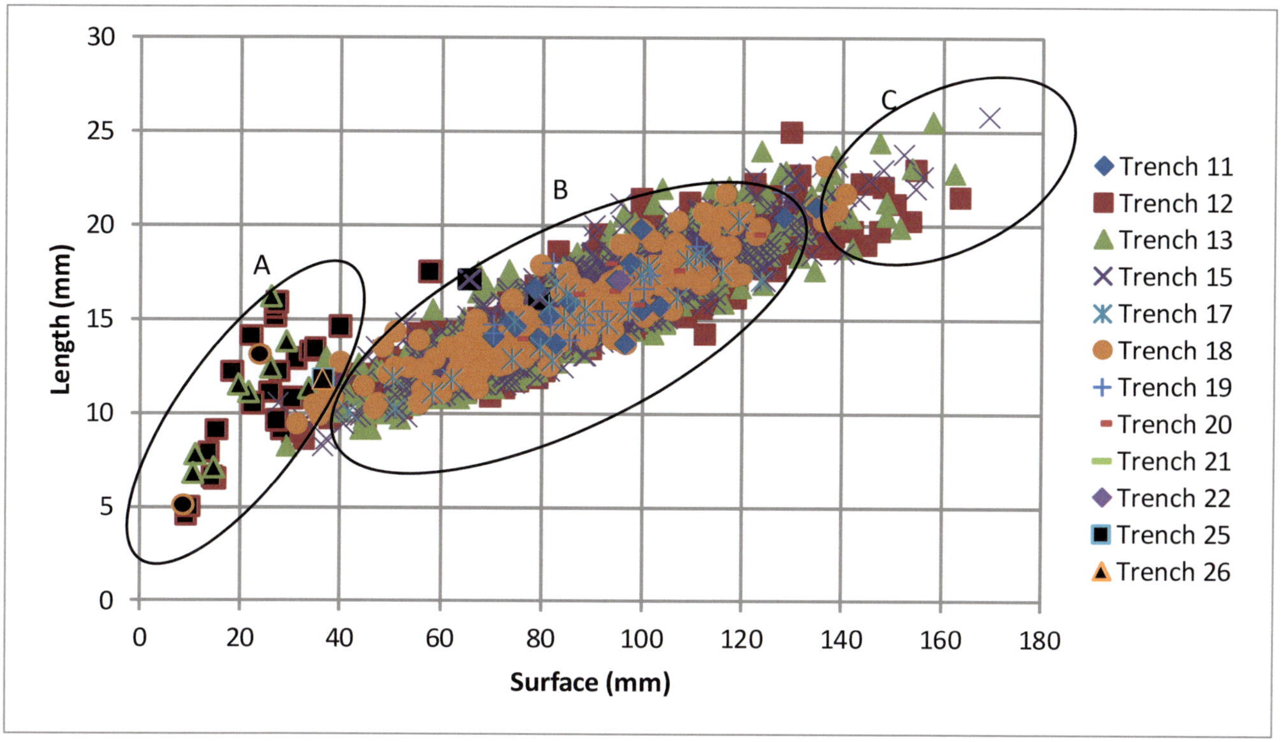

Figure 4. Scatter plot of surface area versus length for the date palm seeds originating from all trenches. Group A: Reduction. Gnawed seeds from the granary (Trench 12), the street (Trench 13), and the western room in building N1 (Trench 15). Group B: Contamination. Mostly, seeds from Trench 18 associated with apparent evidence of contamination. Group C: Mixing. Mostly, the biggest seeds from Trench 12 not associated with apparent evidence of contamination. Measurements: (8.2–)15.65(–25.8) × (3.7–)6.83(–9.6) mm (N = 2730; SD: 2.63 and 0.48).

mostly present a dark brown to reddish colour with poor preservation of the seed coat that may depend on exposure length and burial conditions. They mostly originate from highly mixed trash deposits sealed under levelling layers where no apparent evidence of contamination was found. Therefore, for these specimens, it is tempting to suggest an ancient date, also in consideration of the idea that old date palm seeds tend to be at the top end of the range of measurements (Beech 2003; *contra* Clapham & Stevens 2009). On the other hand, a recent study points to a higher diversity of cultivated forms of *Phoenix* from the beginning of the first century AD (Terral et al. 2012).

Differences in the distribution of such evidence among the trenches give the opportunity to reflect on how the peculiarity of the archaeological context can influence certain taphonomic processes. For example, intense animal activity took place in the granary, most probably in connection with the early disuse of the structure, before any important structural collapse occurred. At this time, maybe because of their seclusion, the bins may have represented preferential feeding places where rodents accumulated the potential food resources available in the surrounding area. Then, with the abandonment of the structure, the bins seem to have attracted various kinds of refuse, conforming to taphonomic behaviour typical of contexts such as pits originally used for storage and then abandoned (Schiffer 1983). Most importantly, architectural collapses and wind-blown deposits occurred to seal these rubbish deposits and, at the same time, safeguarded them from further disturbance (such as contamination with recent material). In Trench 18, where contamination is frequent, the highest number of contaminating items was found in strata between 0.08 and 1.66 m below ground level. In the midden area investigated in the street, rodent activity was intense in the layers beneath the architectural collapses; the character of this highly mixed context suggests it received fairly diverse food waste. Similarly, the richness of food plants suggests that the western room in building N1, once it had been abandoned, was intensively used as a dump, most likely by nearby households. In this context the concentration of plant remains can be explained by disposal patterns combined with the length of surface exposure of plant remains. Disposal patterns may have involved the re-deposition of rubbish that had originally accumulated elsewhere; moreover, successive refuse depositions may have preserved the plant remains

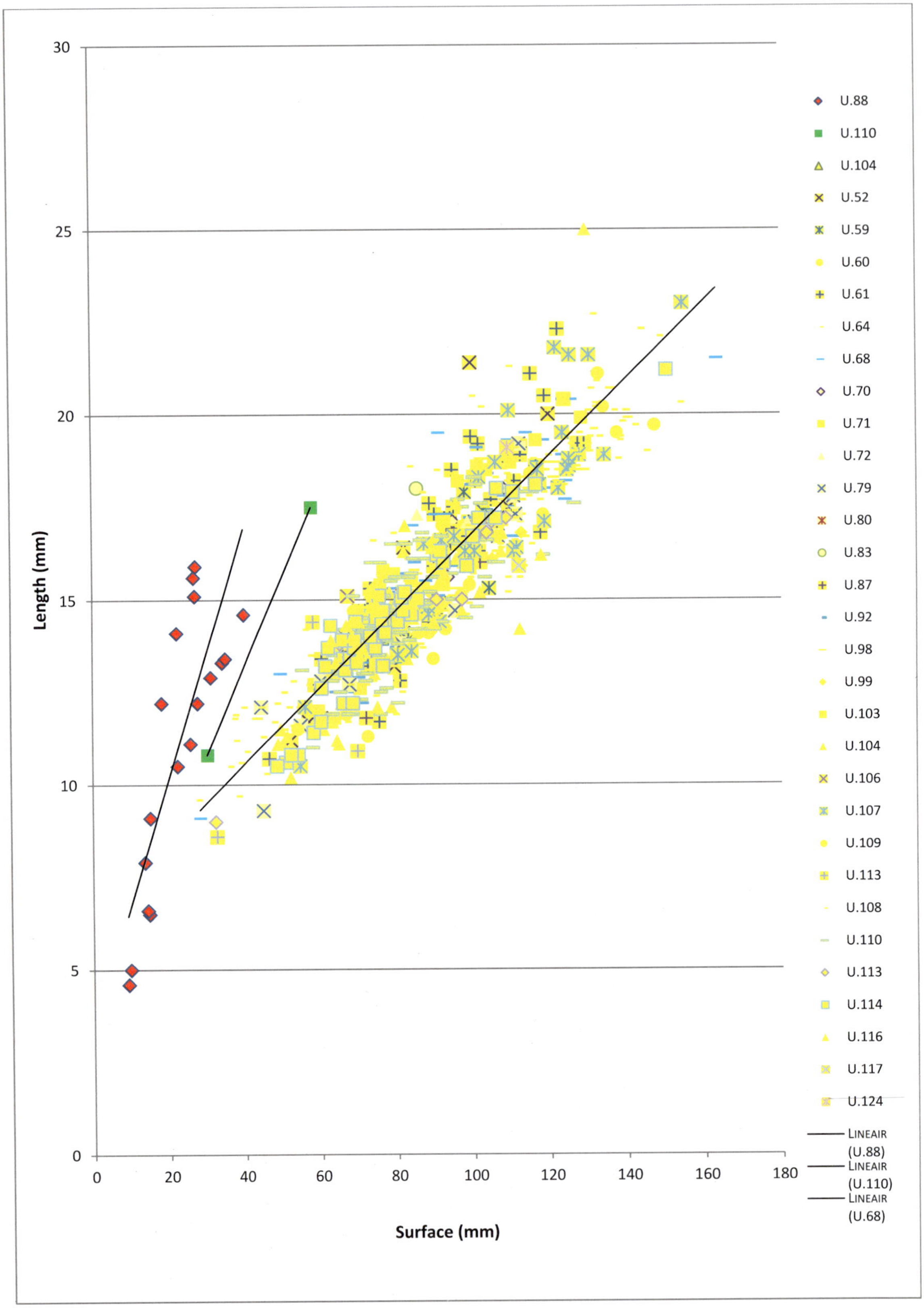

Figure 5. Scatter plot of surface area versus length for the date palm seeds originating from the granary bin (Trench 12). Red diamonds and green squares: gnawed specimens retrieved, respectively, from Unit 88 and Unit 110, most probably representing single depositions. Black dashes: the specimens from Unit 68, which cover the entire range of measurements of the group. Others: all the other specimens from the same deposit.

present in the lower layers from the actions of abiotic as well as biotic factors. From this point of view, a shorter exposure may correlate well to a lower gnawing rate, as suggested by the trend of the gnawed date seeds retrieved in this context.

It is thus possible to tie the groups of specimens defined on the basis of the morphometric analysis to the processes of reduction, contamination, and mixing. More problematic, however, is the use of this type of analysis for distinguishing single depositions. This is particularly true when relatively large numbers of seeds, even from a single unit, differ in size (and shape). For example, the date palm seeds recovered from samples sieved on-site often represented large batches of between 19 and 125 items that were morphologically very diverse. It is clear that in these cases a morphometric analysis would produce a scatter distribution and blur potential single depositions. The morphometric analysis of the only date seeds from the granary bin (Trench 12) may serve as an example (Fig. 5).

The majority (n = 17) of the gnawed seeds present in the deposit originated from a single unit lying on top of the bin floor at the north-eastern corner of the granary courtyard (U.88) (in the red diamonds). Because of their analogous state of preservation, for these specimens an interpretation as single deposition seems reasonable. In the same layer, high proportions of grain kernels of hard wheat (*Triticum turgidum* ssp. *durum* (Desf.) Husn.) and endocarps of olive (*Olea europea*) were retrieved that showed rodent incisor marks. One single deposition may even be suggested for two outliers from U.110 (in green squares), but this remains speculative. The largest group of date seeds (in yellow) shows a scatter distribution, which can be observed even within the same unit. This is, for instance, the case with U.68 (in black dashes), where a large number of specimens (n = 69) covers the entire range of measurement of the group. Such high diversity does not necessarily rule out the possibility that these seeds were actually deposited in one episode. In these cases, the morphometric analysis cannot be carried out in isolation; integration with other botanical and archaeological evidence is imperative.

4. Conclusions

Despite the complexity of the processes that may affect the representativeness of the archaeobotanical archive, an attempt can be made to retrace the main taphonomic processes of dispersal/deposition, mixing/contamination, and reduction. This is especially true when taphonomy is investigated as a potential source of bias in tandem with the archaeological context.

Accounting for taphonomic issues is necessary in order to rectify the bias in the representation of taxa that complicates reconstructing their contribution to the past food economy. For example, in the case of date palm, it is clear that fragmentation, due to rodent-gnawing activity, and contamination may represent major sources of bias. For this reason, it is necessary, first, to discriminate between recent and subfossil remains and, second (and ideally), to numerically adjust the number of fragments and convert them to an estimated number of whole specimens. An attempt has been made to apply the analysis of morphometric features of date palm seeds extrapolated from digital images to see whether it is possible to correlate groups of items with certain taphonomic processes.

The analysis of morphometric features of date palm seeds facilitated separating groups of items and correlating them with the processes of mixing, contamination, and reduction. This separation was possible by taking into account the characteristic features of the seeds, as well as the overall botanical composition and the stratigraphy of the related deposit. The differences observed among and within the trenches can be correlated with the peculiarities of the archaeological context regarded as the basin of deposition. Understanding how and to what extent taphonomic processes are preferentially distributed among different contexts can give a preliminary indication of these contexts' representativeness and thereby guide the sampling strategy.

Much more problematic is the application of morphometric analysis to the identification of single acts of deposition. In this case the autonomy of the method is even more limited and the contribution of other botanical and archaeological evidence becomes indispensable. Date palm seeds have the disadvantage that, normally, a large number of items, very diverse in size and shape, can originate from the same unit. In this respect, groups of items made more uniform by processes affecting seed dimension and shape, such as rodent-gnawing activity, offer a better chance to pinpoint possible single depositions.

The application of digital imaging and measuring techniques to the study of taphonomy discussed above makes it clear that, as in other current fields of botanical research, the use of technology cannot be independent; we still need to rely also on the morphological analysis of plant remains. This includes not only—and most obviously—a correct identification, but also the careful observation of characteristic features that can result from biotic and abiotic activity. This does not solve the bias introduced by missed identifications of items that are particularly small and aberrant in shape, such as the seed fragments of date palm that are leftovers of rodent activity. These items can be especially overlooked if hand-picking is performed in the field by non-specialists. In this case, training in their identification may be of importance in order to prevent biased representation of the species in the hand-picked archaeobotanical samples.

References

Beech M., 2003. Archaeobotanical evidence for early date consumption in the Arabian Gulf. In: Emirates Center for Strategic Studies and Research (ed.), *The Date Palm—From Traditional Resource to Green Wealth.* Emirates Center for Strategic Studies and Research, Abu Dhabi, pp. 11-31.

Beech M. & Shepherd E., 2001. Archaeobotanical evidence for early date consumption on Dalma Island, United Arab Emirates. *Antiquity* 75, pp. 83-89.

Boak A.E.R. & Peterson E.E., 1931. *Karanis, Topographical and Architectural Report of Excavations during the Seasons 1924–1928.* University of Michigan Press, Ann Arbor.

Cappers R.T.J. 2006. *Roman Foodprints at Berenike: Archaeobotanical Evidence of Trade and Subsistence in the Eastern Desert of Egypt.* Cotsen Institute of Archaeology Monograph 55, University of California, Los Angeles.

Cappers R.T.J. & Hamdy R., 2007. Ancient Egyptian plant remains in the Agricultural Museum (Dokki, Cairo). In: Cappers R.T.J. (ed.), *Fields of Change: Progress in African Archaeobotany.* [Proceedings of the 4th International Workshop on African Archaeobotany.] Groningen Archaeological Studies 5. Barkhuis & Groningen University Library, Groningen, pp. 165-214.

Cappers R.T.J. & Neef R., 2012. *Handbook of Plant Palaeoecology.* Barkhuis & Groningen University Library, Groningen.

Cappers R.T.J., Sikking L., Darnell J.C. & Darnell D., 2007. Food supply along the Theban desert roads (Egypt): The Gebel Roma, Wadi el Huôl and Gebel el-Gir caravansary deposits. In: Cappers R.T.J. (ed.), *Fields of Change: Progress in African Archaeobotany.* [Proceedings of the 4th International Workshop on African Archaeobotany.] Groningen Archaeological Studies 5. Barkhuis & Groningen University Library, Groningen, pp. 127-138.

Clapham A.J. & Stevens C.J., 2009. Dates and confused: Does measuring date stones make sense? In: Ikram S. & Dodson A. (eds.), *Beyond the Horizon: Studies in Egyptian Art, Archaeology and History in Honour of Barry J. Kemp.* Publications of the Supreme Council of Antiquities, Cairo.

Clapham A.J. & Stevens C.J., 2012. The plant remains from the Stone Village. In: Stevens A. (ed.), *Akhenaten's Workers: The Amarna Stone Village Survey, 2005–2009.* Vol. 2. The Faunal and Botanical Remains, and Objects. EES Excavation Memoir 100-101. Egypt Exploration Society, London, pp. 15-45.

Cox A. & Van der Veen M., 2008. Changing foodways: Watermelon (*Citrullus lanatus*) consumption in Roman and Islamic Quseir al-Qadim, Egypt. *Vegetation History and Archaeobotany* 17 (Supplement 1), pp. 181-189.

Dennel R., 1972. The interpretation of plant remains: Bulgaria. In: E.S. Higgs (ed.), *Papers in Economic Prehistory.* Cambridge University Press, Cambridge, pp. 149-159.

Dennel R., 1974. Prehistoric crop processing activities. *Journal of Archaeological Science* 1, pp. 275-284.

Dennel R., 1976. The economic importance of plant resources represented on archaeological sites. *Journal of Archaeological Science* 3, pp. 229-247.

Fantone F., 2012. The representativeness of the archaeobotanical archive: Examples from Roman Karanis. Unpublished MA Thesis, University of Groningen, Groningen.

Gasser R.E. & Adams E.C., 1981. Aspects of deterioration of plant remains in archaeological sites: The Walpi Archaeological Project. *Journal of Ethnobiology* 1, pp. 182-192.

Gazda E.K. (ed.), 1983. *Karanis, an Egyptian Town in Roman Times.* Kelsey Museum of Archaeology, University of Michigan Press, Ann Arbor.

Gifford D.P., 1981. Taphonomy and palaeoecology: A critical review of archaeology's sister disciplines. *Advances in Archaeological Method and Theory* 4, pp. 365-438.

Green F.J., 1979. Phosphate mineralization of seeds from archaeological sites. *Journal of Archaeological Science* 6, pp. 279-285.

Hillman G.C., 1981. Reconstruction of crop-husbandry practices from charred remains of crops. In: Mercer R. (ed.), *Farming Practice in British Prehistory.* Edinburgh University Press, Edinburg, pp. 123-162.

Hubbard R.N.L.B. & Clapham A.J., 1992. Quantifying macroscopic plant remains. *Review of Palaeobotany and Palynology* 73, pp. 117-132.

Husselman E.M., 1979. *Karanis Excavations of the University of Michigan in Egypt 1928–1935: Topography and Architecture: A Summary of the Reports of the director, Enoch E. Peterson.* University of Michigan Press, Ann Arbor.

Jones G.E.M. & Halstead P., 1995. Maslins, mixtures and monocrops: On the interpretation of archaeobotanical crop samples of heterogeneous composition. *Journal of Archaeological Science* 22, pp. 103-114.

Miksicek C.H., 1987. Formation and processes of the archaeobotanical record. *Advances in Archaeological Method and Theory* 10, pp. 211-247.

Miller, N., 1996. Seed eaters of the ancient Near East: Human or herbivore? *Current anthropology* 37 (3), pp. 521-528.

Murray M.A., 2000. Fruits, vegetables, pulses and condiments. In: Nicholson P.T. & Shaw I. (eds.), *Ancient Egyptian Materials and Technologies*. Cambridge University Press, Cambridge, pp. 609-655.

Neef R., Cappers R.T.J. & Bekker R.M., 2012. *Digital Atlas of Economic Plants in Archaeology*. Barkhuis, Groningen.

Nesbitt M., 1993. Archaeobotanical evidence for early Dilmun diet at Saar, Bahrain. *Arabian Archaeology and Epigraphy* 4, pp. 20-47.

Newton C., 2004. Plant tempering of Predynastic pisé at Adaïma in Upper Egypt: Building material and taphonomy. *Vegetation History and Archaeobotany* 15, pp. 55-64.

Nowak R.M. & Paradiso J.L., 1983. *Walker's Mammals of the World*. Vol. 1. Johns Hopkins University Press, Baltimore.

Panagiotakopulu E. & Buckland P.C., 2010. The insect remains from Ranefer's house and Grid 12. In: Kemp B. & Stevens A. (eds.), *Busy Lives at Amarna: Excavations in the Main City (Grid 12 and the House of Ranefer, N49.18)*. Vol. I. The Excavations, Architecture and Environmental Remains. EES Excavation Memoir 90. Egypt Exploration Society and Amarna Trust, London, pp. 427-443.

Rathbone D.W., 1990. Villages, land and population in Graeco-Roman Egypt. *Proceedings of the Cambridge Philological Society* 36, pp. 103-142.

Rowley-Conwy P., 1987. Remains of date (*Phoenix dactylifera*) from Failaka, Kuwait. In: Hojlund F. (ed.), *Danish Archaeological Investigations on Failaka, Kuwait: The Second Millennium Settlement*. Vol. 2. The Bronze Age Pottery. Jutland Archaeological Society, Aarhus University Press, Aarhus, pp. 181-183.

Rowley-Conwy P., 1994. Dung, dirt and deposits: Site formation under conditions of near-perfect preservation at Qasr Ibrim, Egyptian Nubia. In: Luff R.M. & Rowley-Conwy P. (eds.), *Whiter Environmental Archaeology*? Oxbow Monograph 38. Oxbow Books, Oxford, pp. 25-32.

Schiffer M.B., 1972. Archaeological context and systemic context. *American Antiquity* 37 (2), pp. 156-165.

Schiffer M.B., 1976. *Behavioural Archaeology*. Academic Press, New York.

Schiffer M.B., 1983. Toward the identification of formation processes. *American Antiquity* 48 (4), pp. 675-706.

Smith W., 2003. *Archaeobotanical Investigations of Agriculture at Late Antique Kôm el-Nana (Tell el-Amarna)*. Egypt Exploration Society, London.

Stevens C.J. & Clapham A.J., 2010.The botanical samples. In: Kemp B. & Stevens A. (eds.), *Busy Lives at Amarna: Excavations in the Main City (Grid 12 and the House of Ranefer, N49.18)*. Vol. I. The Excavations, Architecture and Environmental Remains. EES Excavation Memoir 90. Egypt Exploration Society and Amarna Trust, London, pp. 427-443.

Terral J.F., Newton C., Ivorra S., Gros-Balthazerd M., Tito de Moiras C., Picq S., Tenberg M. & Pintaud J.C., 2012. Insights into the historical biogeography of the date palm (*Phoenix dactylifera* L.) using geometric morphometry of modern and ancient seeds. *Journal of Biogeography* 39, pp. 929-941.

Thanheiser U., 2008. Plant remains from habitation areas in Kellis: Some considerations concerning their accumulation. In: Wiseman M. (ed.), *The Oasis Papers II: Proceedings of the Second International Conference of the Dakhleh Oasis Project*. Dakhleh Oasis Project Monograph 12. Oxbow, Oxford, pp. 141-150.

Van der Veen M., 2007. Formation processes of desiccated and carbonized plant remains: The identification of routine practice. *Journal of Archaeological Science* 34, pp. 968-990.

Van der Veen M., 2011. *Consumption, Trade and Innovation. Exploring the Botanical Remains from the Roman and Islamic Ports at Quseir al-Qadim, Egypt*. Africa Magna Verlag, Frankfurt.

Wendrich W., 2010. *Egyptian Archaeology*. Blackwell Studies in Global Archaeology. Wiley-Blackwell, Oxford.

Willcox G., 2004. Measuring grain size and identifying Near Eastern cereal domestication: Evidence from the Euphrates valley. *Journal of Archaeological Science* 31, pp. 145-150.

Archaeobotanical research and related ethnobotanical observations in the central and southern Sahara

Assunta Florenzano, Eleonora Rattighieri, Isabella Massamba N'siala & Anna Maria Mercuri

Laboratorio di Palinologia e Paleobotanica, Dipartimento di Scienze della Vita
Università di Modena e Reggio Emilia
Viale Caduti in Guerra 127
41121 Modena, Italy

E-mail: assunta.florenzano@unimore.it; annamaria.mercuri@unimore.it

Abstract. This paper presents two case studies demonstrating links between archaeobotany and ethnobotany, completed in the south-western and central Sahara desert from 2004 to 2008. The focus is on evidence of plant use from archaeobotanical records, mainly pollen. Ethnobotanical data are reported from the literature, interviews, and observations of the local cultural knowledge of Tuareg people. The archaeological sites represent a range of contexts, including the open air necropolis of Gobero in Niger and rock shelters and cave sites of the Tadrart Acacus mountains in Libya. Samples taken from burials at Gobero had very low pollen concentrations. However, four Pastoral burials preserved high percentages of pollen grains of caper and myrtle and a significant presence of grasses. Particular funerary behaviours are suggested. In the Tadrart Acacus, the actions of hunter-gatherers and pastoralists are visible in the archaeobotanical record. Remains of food and medicinal plants, such as fruits of *Balanites* and pollen of *Artemisia*, were typically observed in layers corresponding to the early and mid-Holocene. Huge numbers of Poaceae florets and caryopses and diverse fruits, seeds, leaves, and other plant parts accumulated in shelters. Altogether, at least 38 pollen types pertain to plants that were probably used for food and fodder.

Keywords: archaeobotany, ethnobotany, central Sahara, Libya, Niger

1. Introduction

The study of plant–people interactions in modern societies can provide invaluable information to understand aspects of unwritten heritage that has survived to the present day from millennia of traditions. Diverse and interconnected approaches are offered by anthropology; ethnobotany, including ethnopharmacology and prehistoric and historical archaeobotany; linguistic, horticultural, and agrarian sciences; iconography and art history; and sociology (Cunningham 1997; Diamond 2002; di Lernia & Zampetti 2008; Mercuri 2008a; Neuwinger 2000; Okigbo et al. 2008; Zohary & Hopf 2000). The need for interdisciplinary research in this field is urgent because indigenous peoples, their habitats, and traditional lifestyles are under threat (Prance 1991).

In this sense, archaeobotany and ethnobotany are closely related sciences, both dealing with the relationships between humans and plants. The first approach is centred on plant remains from past contexts, while the second mainly investigates traditional knowledge still preserved in living peoples. Archaeological records permit the inference of the cultural value of plants from their nature, quantity, and spatial distribution (e.g. Cappers 2006, 2007; Mercuri 2001; Van der Veen 1999, 2007), while ethnobotanical observations collect data on present-day plant knowledge and processing (e.g. De Natale & Pollio 2012; Martin 1995; Minnis 2000; Schultes & von Reis 1995). Generally speaking, biologists and archaeologists have different approaches that may result in a distinctive, ecological versus economic-centred perspective on interpreting archaeobotanical contexts (Cappers & Neef 2012;

Van der Veen 2011). However, in past contexts economic choices were mainly based on local ecology and landscape. 'Nature' and 'culture' were intermingled, definitely to a greater extent than they are in modern Western thought (Diamond 2002; Mercuri & Sadori 2014).

The uses of many modern plants have their roots in the past; thus ethnobotany can provide reference information to interpret the cultural significance of archaeobotanical data. This is especially useful when archaeological deposits contain significant quantities of food, spice, and medicinal plants (Pearsall 2000; Van der Veen 2011).

In this paper, we interpret archaeobotanical finds using ethnobotanical evidence and report investigations supported by links between archaeobotanical and ethnobotanical data. Although this is an issue that does not have geographical boundaries or chronological limits (e.g. Bandini Mazzanti et al. 2005; Head et al. 2002), the reported examples were chosen from among north-west African contexts, specifically, two case studies that we have investigated in the south-western and central Sahara desert, in Niger and Libya, respectively, from 2004 to 2008 (Fig. 1). The focus is on the evidence of plant use from archaeobotanical records, mainly pollen. In addition, we report ethnobotanical data from the literature and from direct interviews and observations made concerning the local cultural knowledge of Tuareg people.

Systematic ethnobotanical research was not planned as part of the fieldwork due to logistic and organizational limitations (for example, the female gender of botanists can limit the conversation with Tuareg men) and because fieldwork was especially dedicated to archaeological excavations. Nevertheless, we did take notes on conversations and observations conducted during several seasons of fieldwork. Tuareg peoples were interviewed and followed in the recognizing, gathering, use, and preparation of plants for medical or alimentary purposes. Although a structured questionnaire was filled out by the researchers, the basic information was recorded (by IMN) during personal conversations between the researcher and each of the Tuareg informants. Direct questions were, in general, avoided in order to obtain a set of in-depth replies about the culture of the interviewed person.

1.1 Environment and archaeological contexts

The climate of the south-western and central regions of the Sahara is hyperarid, currently extending between the northern 100 mm and the southern 150 mm isohyets (< 100 mm/year; White 1983), with variable and scanty annual amounts of precipitation and with maximum temperatures occurring in summer. The present-day plant cover consists mainly of desert/sub-desert vegetation,

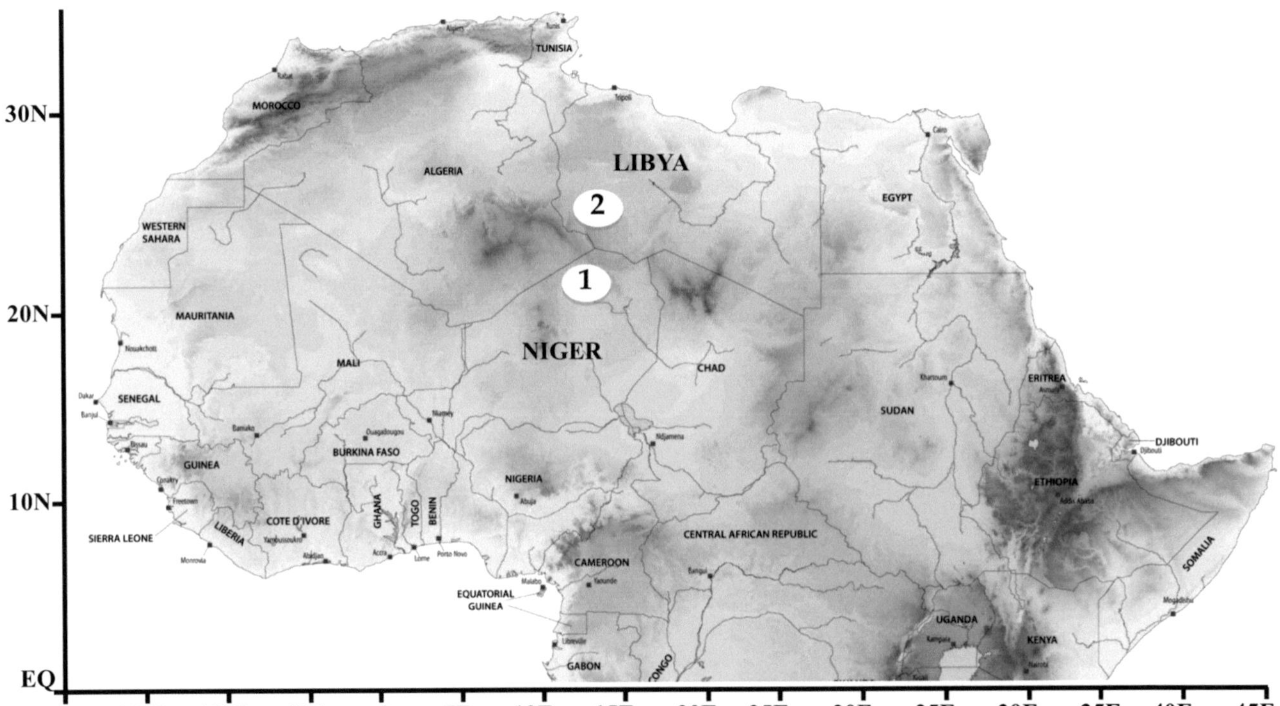

Figure 1. Location of the two case study areas: (1) Gobero, Niger; (2) Tadrart Acacus, Libya.

dominated by *Acacia-Salvadora-Panicum* communities. Wadi vegetation at high elevations is relatively denser than that in the lowlands, and a short-lived achab-flora, in particular, grows in spring after rare precipitation events.

The archaeological sites under discussion present very different contexts, namely, an open air necropolis in Niger and several rock shelters and caves in Libya. These sites were inhabited by people from hunter-gatherer or pastoral cultures. The different chronologies correspond to early or mid-Holocene phases during which the sites are known to have been in use (Cremaschi & di Lernia 1998; di Lernia 1999, 2001; Garcea 2001, 2013; Sereno et al. 2008). The climate was less arid and the vegetation more extensive than it is today. These very different archaeological contexts do, however, have something in common: the people who lived there were challenged by similarly impressive environmental transformations (Mercuri et al. 2011). During times of environmental crisis, and particularly in the more recent periods, populations in both areas faced decreasing water and reduced availability of plant resources that caused famine or the abandonment of their lands.

2. The necropolis of Gobero

In 2006, archaeological fieldwork was conducted at the site of Gobero, Niger, within the framework of a multidisciplinary National Geographic Society project (directed by E.A.A. Garcea and P. Sereno). Gobero is a complex site, with approximately 200 burials spanning five millennia. Considered the earliest and largest Holocene cemetery in the Sahara, this site has opened a new window on our comprehension of the funerary practices, distinctive skeletal anatomy, and health and diet of prehistoric hunter-fisher-gatherers (Garcea 2013; Sereno et al. 2008).

Palynological analyses were carried out with the aim of reconstructing the main features of past environments (climate, ecology, and vegetation) in the southern Sahara in and around the sites (Mercuri et al. 2013). Together with geostratigraphical, faunal, anthropological, and archaeological records, these sites provide a chronicle of episodic human occupation of the Sahara under early and mid-Holocene climatic oscillations (Sereno et al. 2008). The pollen samples were collected from a short sequence in a desiccated lake and from 23 burials attributed to the Pre-Pastoral and Pastoral cultural phases. Pollen from exposed skeletons was carefully collected immediately after the superficial soil layer had been removed, in order to avoid contamination from modern pollen rain. The sampling points were selected near the hands and head, in the area of the stomach, and beneath the skeletons, based on the potential of recovering remains of flowers and plant deposition (Fig. 2).

Although the majority of samples did not yield preserved pollen, about one third provided data that were useful to reconstruct plant landscapes. Pollen was obtained from a total of 14 burials (for detailed analyses of all pollen samples, see Mercuri et al. [2013]; for a discussion on the interpretation of pollen spectra from archaeological contexts, see also Mercuri [2008b] and Mercuri & Garcea [2007]).

According to Giraudi & Mercuri (2013), lacustrine and arid phases from the Gobero basin correlate with an early Holocene humid phase that was interrupted by an 8200 cal BP dry spell, when arid conditions were observed in many regions, including the Mediterranean, the Sahel, and subequatorial Africa (Gasse 2000; Mercuri et al. 2011, and references therein). This was followed by a major dry phase at around 6000 cal BP, with a shift toward more permanent aridity. The progressive drying of the Saharan region occurred in response to a weakened insulation force of the African monsoon and abrupt hydrological change in the aquatic ecosystem of the Chad basin (Kröpelin et al. 2008). Several climatic oscillations occurred in the region, but for the most part, the Sahelian savannah was more widespread than the desert vegetation near Gobero. Therefore, the local environment was more suitable for human life than were other areas of the Sahara (Mercuri et al. 2013).

The samples from burials demonstrated very low pollen concentrations that did not permit them to be considered unequivocal indicators of in-situ plant accumulation. In fact, pollen concentration—in the air and in sediments—depends on both plant phenology and productivity, and the quantity found in sediments also depends on the taphonomy and agent of pollen transport (wind, water, or humans). Human actions affect pollen spectra, determining variations in pollen concentrations—for example, trampling or making fires destroyed pollen in sediments (as in hearth and ash samples), whereas bringing plants to the site concentrated it (as in dung or organic sand samples). Human transport may be hypothesized when high pollen concentrations are found in archaeological layers (Mercuri 2008b) (Fig. 3). In the case of the burials of Gobero, therefore, within the multidisciplinary research, other elements concurred with our interpretation, namely, the point of sampling and the state of preservation of microscopic remains. In addition to being able to perform palaeoenvironmental reconstructions, we were able to infer some particular forms of plant use and relevant funerary behaviour from samples taken from four

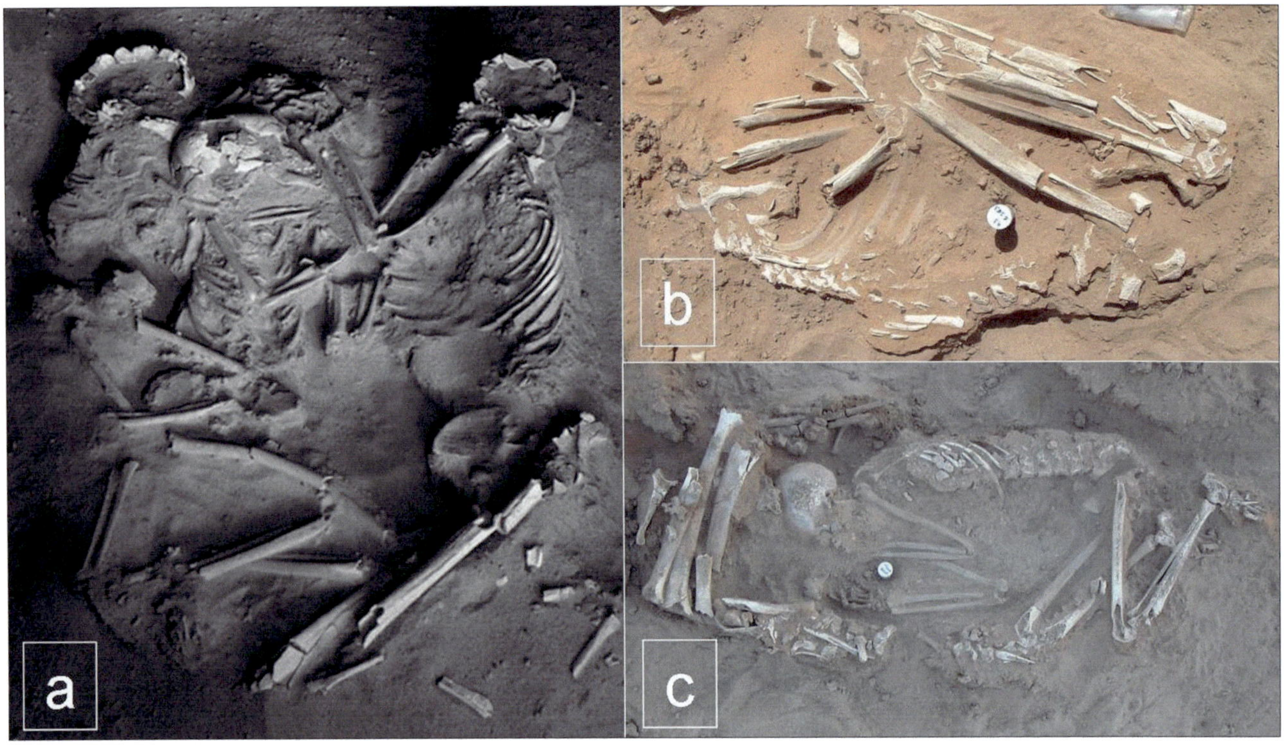

Figure 2. Three burials at Gobero that were studied for pollen and reported on in this paper: (a) the tripple burial G1B8-10; (b) G3B29; (c) G3B6. In figures b and c, the sampling points are visible. (Modified from Mercuri et al. 2013.)

Pastoral burials, two of which were radiometrically dated (Sereno et al. 2008).

2.1 Burial G1B18: Deposit of a cushion under the head of a male

Many graminoid-type phytoliths (silicate inclusions of plant epidermids) were found in one sample (BT-P10 4a) collected under the cranium of a male (Fig. 3). Siliceous phytoliths are normally destroyed by the hydrofluoric acid routinely used in pollen extraction; their recovery in this sample was, therefore, anomalous. Probably the phytoliths were so abundant in the original sample that some specimens survived the processing. Accordingly, the pollen spectrum was dominated by Poaceae–wild grass group (74%) and small amounts of Cyperaceae, *Cornulaca monacantha*-type, and *Boerhavia*-type pollen (≥ 2%). In the same sample, many microscopic fragments of plant tissue and charcoal particles were observed. This uncommonly high presence of phytoliths associated with pollen of grasses under a cranium was interpreted to suggest that a cushion of plants may have been placed under the head of the dead person. Possibly a more extensive layer of plants was also placed under the body as bedding, but no traces of plant tissue were observed in the other four samples taken from this burial.

2.2 Burial G3B29: Poaceae pollen as possible evidence of cereal food

One pollen sample was studied from the soil around the abdomen of this burial (BT-P16 6) (Figs. 2b and 3). It showed a concentration that in this particular context can be considered very high, of about 2800 p/g, which is six times higher than that found in all the other samples from burials at Gobero. Poaceae pollen again largely predominated (74%). This high proportion of grasses is very similar to the previous example. However, it is known that similar spectra may have different origins and that data interpretation is tightly linked to archaeological context. Therefore, in the palynology of archaeological sites, the same percentage may have different significance. In this case, Poaceae include the ≥ 40 μm pollen type (Fig. 4), a large pollen type that is not transported long distances from its source. This pollen is well represented in the spectrum (3.6%), suggesting some ingestion of the source plants. Pollen may be trapped in the glumes after the flowers have matured into fruits, and if ingested, it

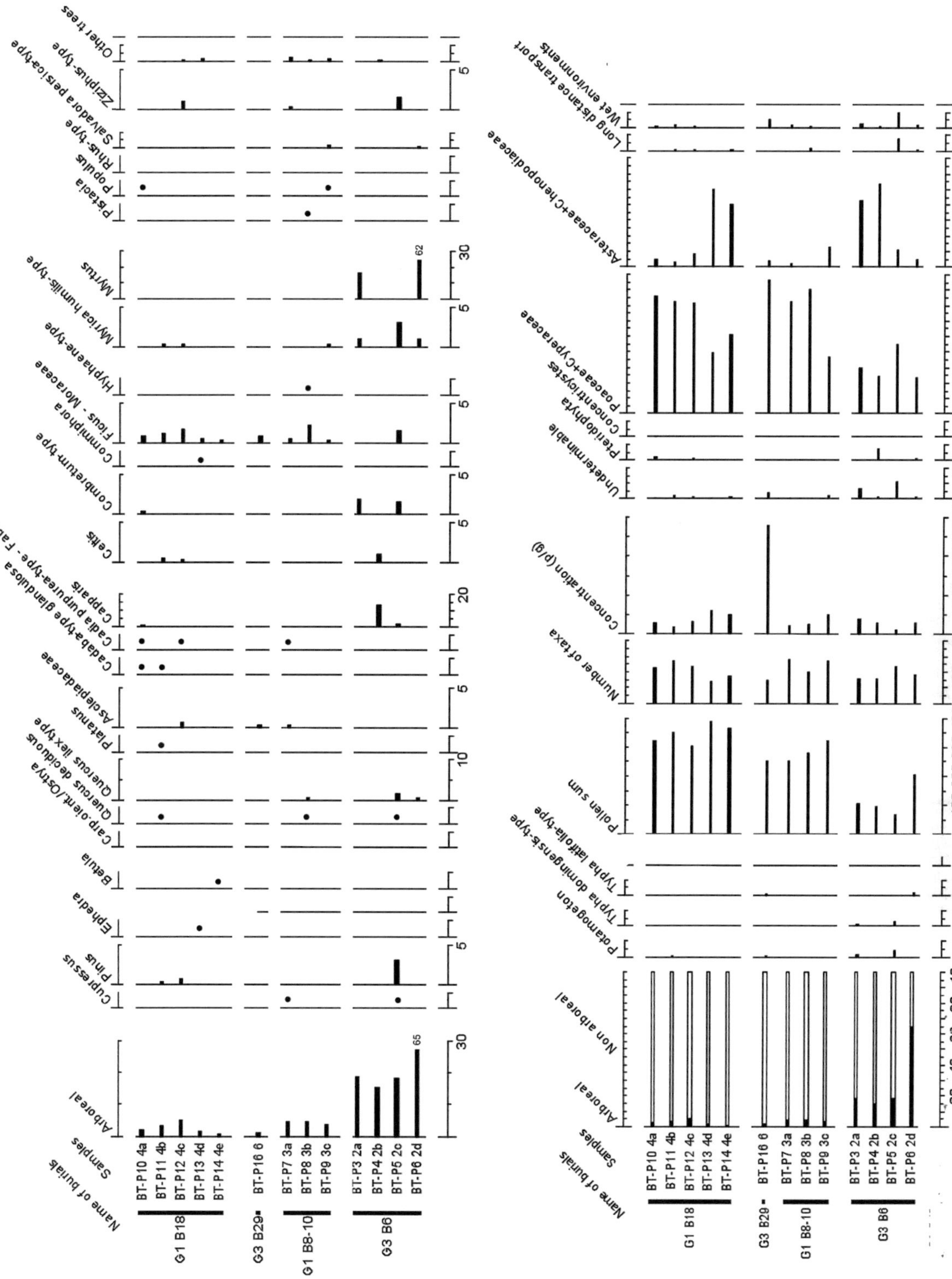

Figure 3. Percentage pollen spectra of woody plants, aquatics, concentrations, and selected totals from the four burials at Gobero discussed in the text. Pollen spectra show low trees and shrubs (arboreal pollen), except for *Myrtus* and *Capparis* pollen from burial G3B6. The hygrophilous *Ficus* is the most common tree represented in pollen spectra; herbaceous plants from wet habitats were also found (e.g. *Potamogeton* and *Typha*), suggesting some water availability in the area. The grassland with Poaceae and Cyperaceae are more common in spectra than are xerophilous taxa, such as Asteraceae and Chenopodiaceae.

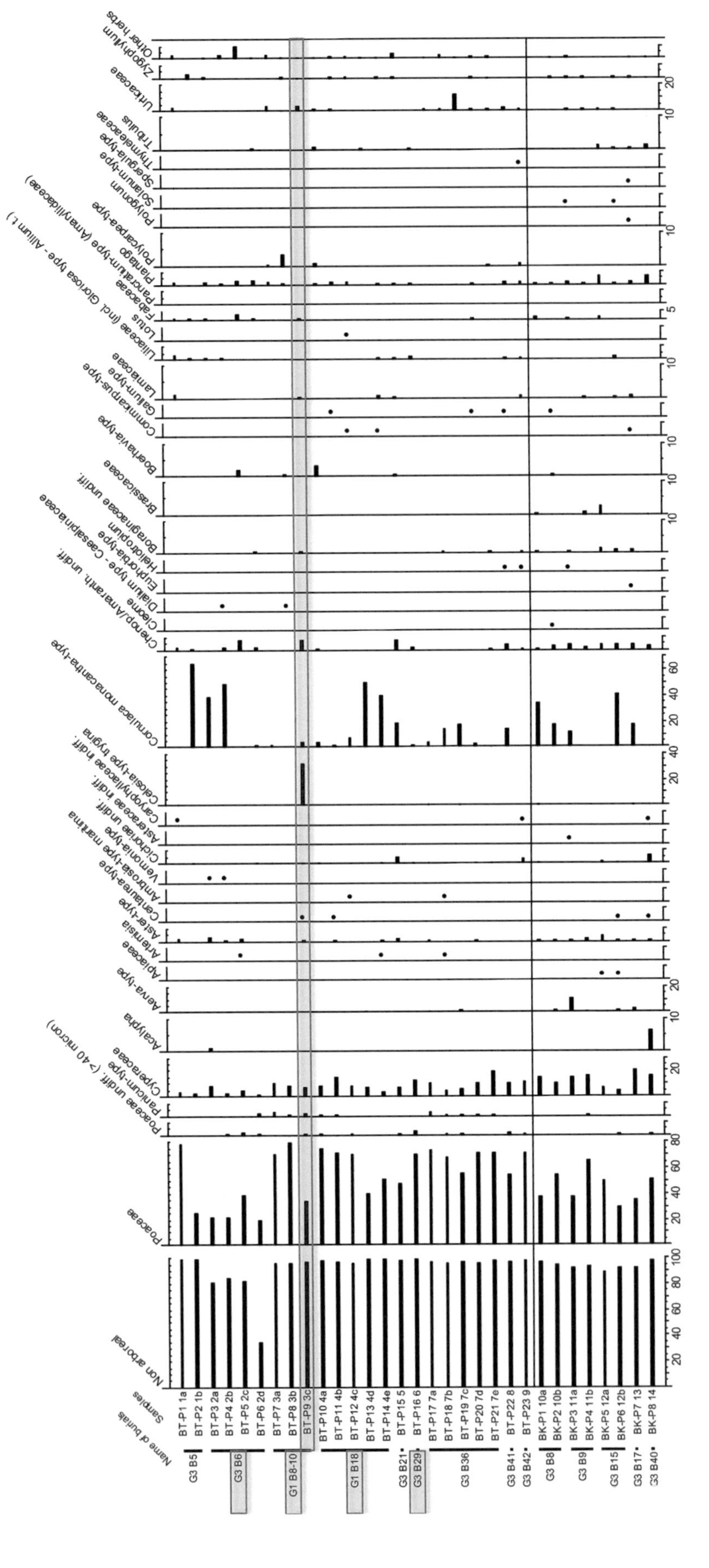

Figure 4. Percentage pollen spectra of non-arboreal plants from the 14 burials at Gobero where samples contain pollen. Grey lines mark the four burials discussed in the text and the sample with exclusive presence of *Celosia*-type *trygina*. (Modified from Mercuri et al. 2013.)

survives the passage through the digestive apparatus (Bottema 1992; Faegri & Iversen 1989). The ≥ 40 µm pollen corresponds to several genera of grasses and several wild cereals, such as the Paniceae *Panicum*, *Echinochloa*, and *Setaria* (Mercuri 1999, 2008b). These wild millets have been harvested for food in the Sahara and the Sahel for millennia (see below). Therefore, this sample may constitute evidence that the last food of this person was based on wild cereals.

2.3 Burials G1B8–10: Deposit of flowers between the arms (about 5300 cal BP)

A triple burial of an adult female and two juveniles showed an intimate co-mingling of arms, hands, and legs (Sereno et al. 2008) (Fig. 2a). Immature pollen grains of *Celosia*-type *trigyna* (Amaranthaceae; Maley 1972) still held within the anther, as well as many accumulations of pollen, were detected in the underlying sediment. This characteristic pollen was found only in this sample at the site of Gobero (BT-P9 3c) (Fig. 4). Poaceae (34%) and *Celosia*-type *trigyna* (32%) monads prevailed in pollen spectra.

Celosia-type *trigyna* includes species of Amaranthaceae with large, sometimes colourful inflorescences. Today, many varieties are cultivated as ornamentals or dye plants. *Celosia trigyna* L. is used in traditional medicine in Niger (Harouna et al. 1996) and in central African countries. For example, it is used in cases of strong fever; as a vermifuge in helminthiasis, mycosis, or eyes diseases; and to cure snake- or scorpion-bites. It may be an occasional food complement in Africa (Adam et al. 1972; Adjanohoun et al. 1988; Société Française d'Ethnopharmacologie 1990). Leaves and stems are soaked to cure brucellosis in domestic animals (Byavu et al. 2000).

2.4 Burial G3B6: Deposit of flowers of myrtle and caper in the burial of a female (about 6000 cal BP)

Two samples taken near the hands and under the cranium of a female burial (Fig. 2c) showed significant percentages of *Capparis* pollen (13% and 2%, respectively). The same pollen was absent from the other two samples (Fig. 3). *Capparis* includes species with beautiful and fragrant flowers that are edible. *Capparis spinosa* L. (caper bush) is still present today in Saharan wadi environments (Fig. 5). The different parts of the plant have many medicinal effects (for a list of chemical compounds and their biological activity, including medicinal properties, see, for example, Mishra et al. [2007]). For example, an infusion of stems and root bark is an antidiarrhoeic and febrifuge; flower buds and roots are useful as a renal disinfectant, diuretic, tonic for arteriosclerosis, and diuretic; and fresh fruits are consumed as an antiscorbutic (Boulos 1983).

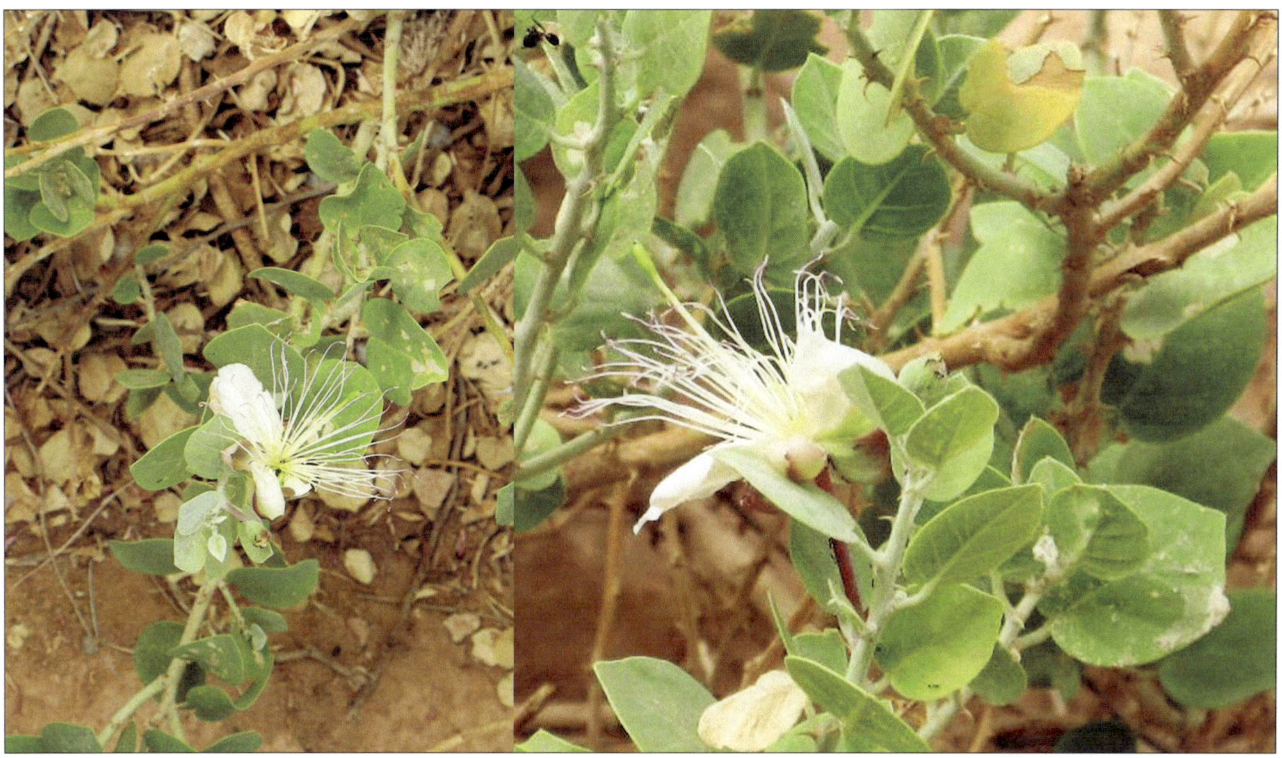

Figure 5. Plant of *Capparis spinosa* observed in the Tadrart Acacus mountains.

Also, the beverage obtained from the maceration of flowers and fruits is used to treat rheumatism (Jaouad et al. 2003). Our Tuareg informants reported that people collect the beautiful and aromatic pink flowers to eat, while the whole plant, mashed with a little quantity of water, is used to cure skin diseases and insect bites.

Soil samples taken under the right foot and near the cranium showed a significant presence of *Myrtus*, a pollen that was exclusive to and uncommonly high in these samples (62% and 16%, respectively) (Fig. 3). *Myrtus nivellei* Batt. et Trab., an endemic species in the Sahara, is closely related to the Mediterranean *M. communis* L. myrtle shrubs that currently grow in Saharan mountains above 1500 m, as, for example, in the more northerly summits of the Aïr (White 1983). This suggests that pastoralists had to travel far from the Gobero site to collect myrtle on the mountains, in order to place it with other scented flowers to decorate the body of this female (Mercuri et al. 2013).

In Mediterranean mythology, myrtle was the emblem of fertility and purity, used in wedding bouquets. When someone died in ancient Greece, houses were decorated with crowns of myrtle and bay. Dionysus went to the goddess of the underworld, Persephone, and offered her myrtle to release his mother from death (Cattabiani 1998). It is the chemical properties of this plant—its basic biological traits—that may have been attractive features for humans, from the mid-Holocene to historical to contemporary times. Myrtle is still an important plant in traditional medicine, particularly in the pharmacopoeia of the Tassili n'Ajjer, where it is used to heal intestinal diseases, fever, and diabetes (Hammiche & Maiza 2006). We observed that leaves and fruits of *Myrtus nivellei* are still used as traditional food and medicine by Tuareg people; for example, tea with myrtle was drunk by one of the authors in 1992 (AMM; mission in Tadrart Acacus directed by Fabrizio Mori), and additional observations and interviews were made in later missions (by IMN). Leaves are dried and reduced to a fine dust that is added to fresh butter made with goat's milk. It helps to preserve the butter, at the same time enhancing the aroma and flavour. Even today, myrtle is collected from wild shrubs growing in the mountains. Today, thanks to modern wells that were mechanically excavated, modern Tuareg people have access to sufficient amounts of water in their desert settlements, but they rarely decide to transplant useful and medicinal plants from their wild habitats and cultivate them near their dwellings.

3. The Tadrart Acacus rock shelters and caves

Since the 1990s, archaeological and multidisciplinary research carried out by the Italian–Libyan Archaeological Mission in the Acacus and Messak, in south-western Libya, has included archaeobotanical analyses. Many rock shelters and caves located along the Wadi Teshuinat and surrounding wadis in the Tadrart Acacus preserved deposits that, in addition to containing valuable archaeological materials, were rich in pollen and desiccated plant remains (Cremaschi & di Lernia 1998) (Fig. 6). In the

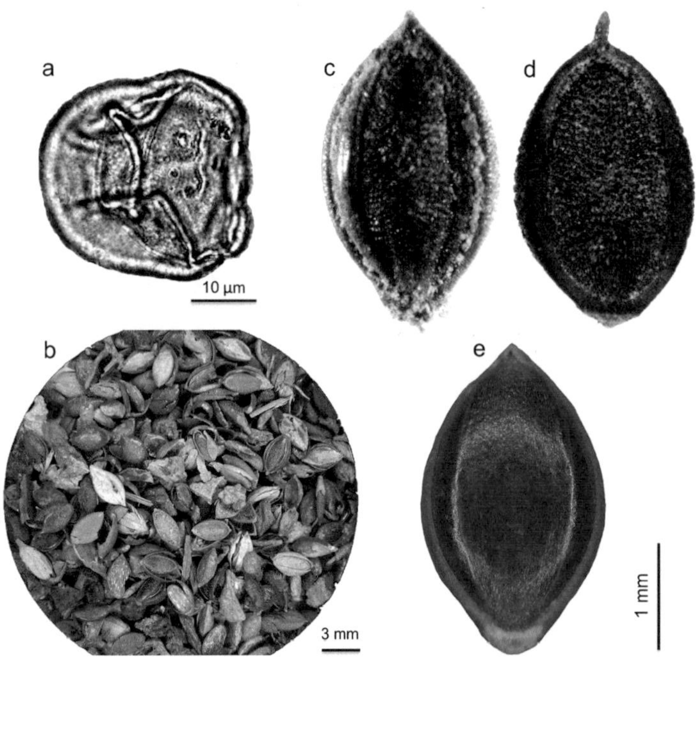

Figure 6. Plant remains of grasses from the archaeological sites of the Tadrart Acacus mountains (Mercuri 1999; Mercuri & Garcea 2007; Olmi et al. 2012): (a) Poaceae pollen > 40 µm from Uan Afuda; (b) assemblage of florets, caryopses, and glumes from Uan Telocat; (c) *Brachiaria* and (d) *Urochloa* from Uan Tabu; (e) *Echinochloa* from Takarkori.

framework of integrated studies, pollen and plant remain data provided the basis for reconstructing vegetation history resulting from climate oscillations and human influence in this region of the central Sahara (Cremaschi & Zerboni 2009; see Mercuri 2008c for detailed reconstruction of vegetational transformations under climate oscillations during the Holocene).

Since the wet and cool phase occurring at the beginning of the early Holocene, approximately 10 800–10 250 cal BP, the region has experienced noticeable oscillations in humidity and temperature, while the plant cover has remained fairly continuous in Saharan regions (Cremaschi & Zerboni 2010; Giraudi et al. 2013; Hoelzmann et al. 2010; Lézine 2009; Lézine et al. 2011) (Fig. 7).

In the Tadrart Acacus, the action of hunter-gatherers is visible in archaeological sites where huge numbers of caryopses and diverse fruits, seeds, leaves, and other plant parts were accumulated in shelters (di Lernia 2002; Mercuri 1999). During the mid-Holocene, the grassland vegetation that was prevalent in the early Holocene shifted southward and the landscape was increasingly colonized by xerophilous vegetation. When pastoralists occupied the area, archaeological deposits preserved layers of dung and coprolites of goats and sheep (di Lernia 2001). The seasonality caused phases of extreme aridity when vegetation cover became discontinuous and lake levels declined. After thousands of years of grazing, animal pressure would have played an important role in shaping the landscape, contributing to reduced plant cover, increased shrublands and depleted soils (Mercuri 2008c). These factors enhanced the effects of the increasing aridity.

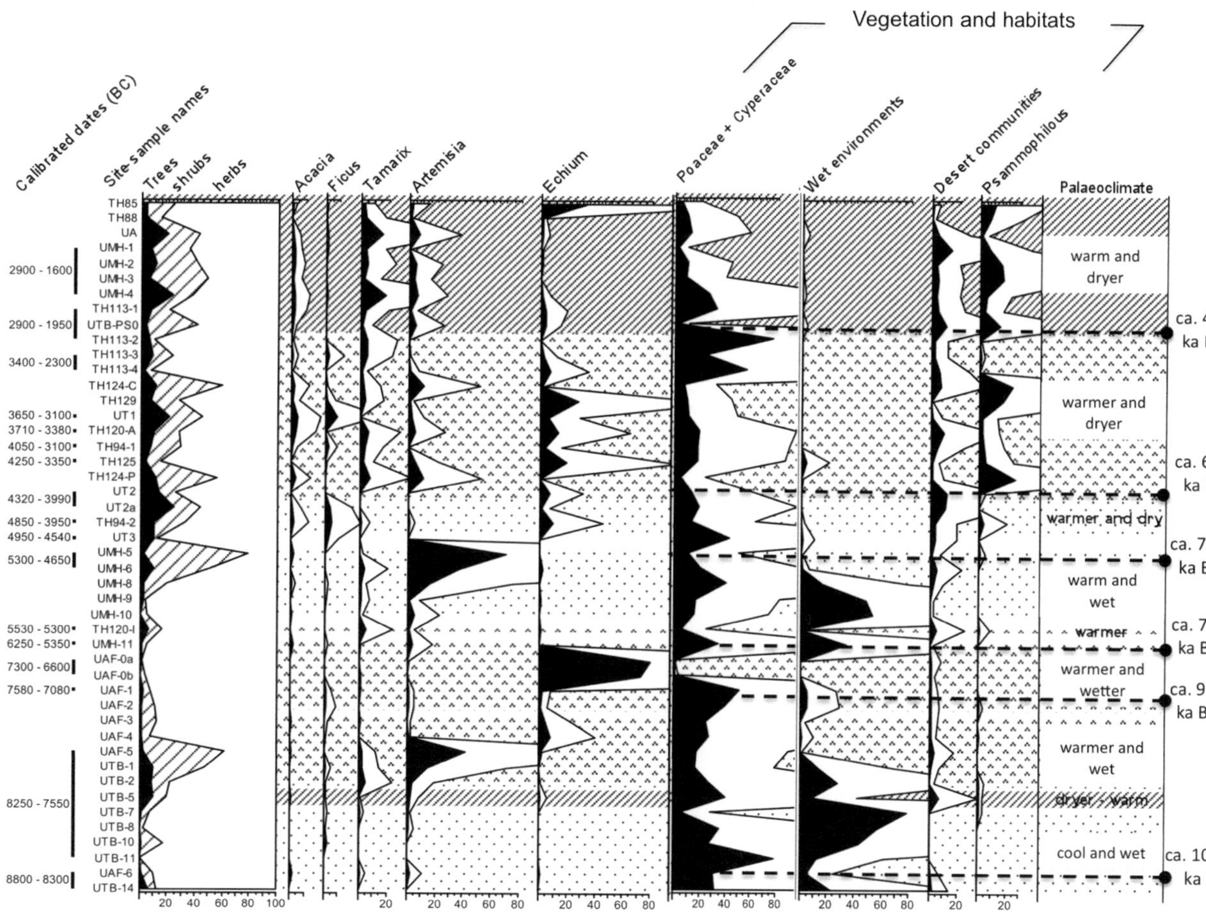

Figure 7. Percentage pollen spectra of selected taxa and totals from 13 sites studied in the Tadrart Acacus mountains from early and mid-Holocene deposits (Wadi Teshuinat). The pollen show the main features of spectra from these arid lands, such as low percentages of trees, among which *Acacia*, *Ficus*, and *Tamarix* are common. The pollen curves of *Artemisia* and *Echium* have sharp peaks that are evidence of human transport of these plants into the rock shelters. Pollen of wet environments are better represented in the earlier phase (before c. 7000 BP), while pollen of desert communities and psammophilous plants increase after this date. This is an effect of increasing dryness in the second part of the Holocene. (Modified from Mercuri 2008c.)

Table 1. Pollen records of plants used for food (A), grazing/fodder (gz), colour (col), medicine (med), and toxins (tox) found in archaeological deposits in rock shelters and caves of the Tadrart Acacus mountains.

Site	Chronology	*Acacia* cf. *tortilis*	*Aerva*-type *javanica*	*Artemisia*	*Asteriscus/Bellis* type	*Astragalus*	*Balanites* cf. *aegyptiaca*	cf. *Boerhavia*	*Boscia*	*Calligonum*	*Capparis*	*Cornulaca* cf. *monacantha*	cf. *Crotalaria*	Cyperaceae	*Echium*	*Euphorbia*	*Fagonia*	*Faidherbia*-type *albida*	*Ficus*	*Helianthemum*
		A - gz - col	A - gz	A - med - gz	gz	gz	A - med	A	A	gz	A - med	gz - med	gz - tox/ med	A - gz	gz	gz	gz - med	gz	A	gz
TH85	Late Pastoral	+		+					+		+	+		+	+	+			+	
TH88	Late Pastoral			+						+	+			+	+					
In Taharin (TH94)	Middle Pastoral	+	+	+						+	+			+	+	+			+	
Immerca (TH113)	Middle and Late Pastoral	+	+	+	+					+	+	+		+	+	+			+	
TH120	Middle and Late Pastoral	+	+	+							+			+	+				+	
TH124	Middle and Late Pastoral	+	+	+	+					+	+	+		+	+	+			+	
TH125	Middle and Late Pastoral		+	+	+						+			+	+				+	+
TH129	Middle and Late Pastoral	+		+						+	+			+	+				+	
Uan Amil	Early, Middle, and Late Pastoral	+	+	+						+	+	+		+	+				+	
Uan Telocat (TH3)	Early, Middle, and Late Pastoral	+	+	+	+				+	+	+			+	+				+	
Uan Muhuggiag (UMH)	Early, Middle, and Late Pastoral	+	+	+	+				+	+	+	+		+	+	+	+	+	+	+
Uan Afuda (UAF)	Late Acacus (Mesolitic and Epipaleolitic hunter-gatherers)	+	+	+	+	+	+	+	+		+	+	+	+	+	+	+	+	+	+
Uan Tabu (UTB)	Late Pastoral	+		+	+					+	+			+	+					
	Late Acacus (Mesolitic and Epipaleolitic hunter-gatherers)		+	+	+						+			+	+	+			+	

Site	Chronology	*Juncus*	*Lavandula* type	*Mentha* type	*Maerua*	*Moltkiopsis* cf. *ciliata*	*Neurada* cf. *procumbens*	*Panicum* type	*Phoenix* cf. *dactylifera*	*Pistacia*	*Plantago*	Poaceae undiff.	*Pulicaria*	*Rhus*	*Rumex*	*Tribulus* cf. *terrestris*	*Trichodesma*	*Typha*	cf. *Zilla*	*Ziziphus*
		A	A - gz	A - gz	A - med	gz	gz	A - gz	A	A - med	gz	gz	gz	A - col	A - gz	A - med	gz, tox/ med, col	A - int	gz	A - gz
TH85	Late Pastoral					+		+				+	+	+			+		+	+
TH88	Late Pastoral					+		+				+	+		+		+			+
In Taharin (TH94)	Middle Pastoral				+	+		+				+	+	+		+	+	+	+	
Immerca (TH113)	Middle and Late Pastoral				+	+	+	+				+	+	+	+	+	+	+	+	
TH120	Middle and Late Pastoral					+		+			+	+	+	+	+	+	+	+	+	+
TH124	Middle and Late Pastoral				+	+		+			+	+	+	+		+	+	+	+	
TH125	Middle and Late Pastoral					+		+				+	+					+		
TH129	Middle and Late Pastoral					+		+				+	+	+	+					
Uan Amil	Early, Middle, and Late Pastoral										+	+	+	+		+	+	+	+	+
Uan Telocat (TH3)	Early, Middle, and Late Pastoral				+	+	+	+			+	+	+	+	+	+	+	+		+
Uan Muhuggiag (UMH)	Early, Middle, and Late Pastoral				+	+	+	+	+		+	+	+	+	+	+		+	+	+
Uan Afuda (UAF)	Late Acacus (Mesolitic and Epipaleolitic hunter-gatherers)	+	+	+	+	+	+	+		+	+	+	+	+		+	+	+	+	+
Uan Tabu (UTB)	Late Pastoral				+	+		+				+	+		+	+	+	+	+	
	Late Acacus (Mesolitic and Epipaleolitic hunter-gatherers)	+			+	+	+	+		+	+	+	+	+	+	+	+	+	+	

Table 2. Examples of plants that are known to be used, based on our ethnobotanical observations and on data from the literature (Corti 1942; Garba 2000; Luxereau 1994; Smith 1980). Common names—where known—are reported in Tamashek, currently spoken in Libya, Mauritania, and Niger, and in Haoussa, spoken in Niger and Mali. The generic word *ishiban* means several wild cereals (Burnus 1992-93).

Scientific name	Name in Tamashek language	Plant part(s) used						Main use
		Trunk, bark	Roots or hypogean parts	Leaves	Flowers	Fruits	Seeds	
Acacia ehrenbergiana Hayne	Tamat	+					+	medicinal
Acacia tortilis (Forssk.) Hayne ssp. *raddiana* (Savi) Brenan	Abser	+				+	+	pasture/fodder, timber
Balanites aegyptiaca Del.	Tebôrak			+	+	+	+	perfume, medicine, pasture/fodder
Borassus aethiopum Mart.		+				+	+	timber, food
Boscia senegalensis (Pers.) Lam. ex Poir.	Tadanti			+		+	+	food, medicinal, timber
Brachiaria lata (Schumach.) C.E.Hubb	Ishiban					+	+	food
Cenchrus biflorus Roxb.	Wajag					+	+	food
Citrullus colocynthis (L.) Schrad.	Alcot, Tagellit, Alke, Alikidd					+	+	pasture/fodder, fuel
Echinochloa colona L.	Ishiban					+	+	food
Hyphaene thebaïca (L.) Mart.	Tagait				+	+	+	food, ornamental, basketry, medicinal, timber
Panicum laetum Kunth	Asghal					+	+	food
Panicum miliaceum L.						+	+	food
Panicum turgidum Forssk.	Afezu, Afasu, Afaso					+	+	food
Pennisetum dichotomum Forssk.	Tehaua, Tarfave, Talemfezzout						+	food
Setaria sp.; *S. pumila* (Poiret) Roemer & Schultes; *S. italica* (L.) Beauv.						+	+	food
Sorghum aethiopicum (Hackel) Rupr. ex Stapf; *S. bicolor* (L.) Moench	Abogha					+	+	food
Vitex doniana Sweet				+		+		timber
Ziziphus spina-christi (L.) Desf.	Abaga, Labacet, Tabakat					+	+	food

3.1 *Seeds/fruits and pollen as ethnobotanical evidence: The examples of* Balanites *and* Artemisia

In these archaeological sites, high percentages of specific pollen grains and concentrations of seeds/fruits point to the accumulation and handling of plants by humans (Mercuri 1999; Olmi et al. 2007; Olmi et al. 2011). Remains of food and medicinal plants were typically observed in layers corresponding to hunter-gatherer and early pastoral occupations. In a review of palynological studies from 13 archaeological sites located in the Tadrart Acacus, Mercuri et al. (2006) reported evidence of plants related to food consumption by humans and domesticated animals. At least 38 pollen types in the floristic list include plants used for food and fodder (Tables 1 and 2). Interestingly, besides grasses (see below), other taxa are recurrent in the archaeobotanical record of many rock shelters and caves. Sugar date and ragweed are two interesting examples.

Balanites aegyptiaca (L.) Delile (sugar date) was encountered as whole or broken fruits (drupes), sometimes abundant, and observed with holes that were probably drilled into the seeds by insects (for other examples, see Cappers 2006; Cappers & Neef 2012; Van der Veen 1995; Wasylikowa 1992; and Mercuri et al. unpublished data). The species has entomophilous pollination, and its pollen naturally occurs in small quantities, but it was nevertheless found in the Uan Afuda cave deposit.

The species is an example of a multipurpose plant, used for food, fodder, oil, and medicine in both African and Asian countries (Gour & Kant 2012). Oil is extracted from its fruits, which are rich in saturated fatty acids, and this oil is ingested to treat headache, spleen, and liver disorders and to improve lactation in traditional medicine. In the past, balanos oil was used in the production of perfume; it was also one of the ingredients of *stacte*, a myrrh oil (Cappers 2006). The fruits contain steroids (saponins, sapogenins, diosgenins) (Ndoye et al. 2004). In many African countries, the aqueous extracts of the fruit are used as a folk remedy for diabetes mellitus (Cappers 2006). In West African traditional medicine, a decoction of the bark is used as an antidote to poisons and to induce abortion. The kernel powder has anthelminthic and molluscicidal properties used to treat goats (Hamidou et al. 2002; Iwu 1993).

Although the (palaeo)ethnobotanical interpretation may be obvious when macroremains are common and observed in large amounts in pits, houses, and other contexts of archaeological sites, the pollen evidence is unquestionable—at least for a palynologist—when a high concentration and percentage of one taxon is found in anthropogenic layers (Mercuri 2008b). In these cases, pollen is the result of the deposition of one selected plant, collected during the blooming period or with pollen trapped in leaves and other parts. This is, for example, the case with the pollen of *Artemisia*, found as monads and anthers, in a few samples from Uan Afuda and Uan Muhuggiag. Macroremains of this genus were not observed in samples from these sites, but the recovery of *Artemisia* pollen from these sites is evidence that the plant grew in the region and was used by people. Samples showing very high percentages of this pollen are, in fact, the result of accumulation of the plant in-situ (Fig. 7). Moreover, the presence of this pollen in human coprolites suggested the ingestion of food and/or beverages at Uan Afuda (Mercuri 1999).

Some species of *Artemisia* are collected for firewood and are used as an aromatic flavouring and spice (Corti 1942). Known since ancient Egyptian times, *Artemisia* has had many uses as a medicinal plant, as a febrifuge, vermifuge, and antiseptic (Bekalo et al. 2009; Bora & Sharma 2011; Erasto et al. 2005; McGovern et al. 2010), and it contains several substances that are present in different proportions throughout the year. According to De Natale and Pollio (2012), even today in Libya and North Africa, several species have traditional uses. *Artemisia arborescens* L. is an anthelmintic, aperitive, diuretic, emmenagogue, and abortive, and helps to treat colds and vertigo (Bellakhdar et al. 1991). The powdered leaves are externally applied to skin infections and to prevent wrinkles (Merzouki et al. 2000). In Tassilli N'ajjer (southern Algeria), aerial parts and flowers of *Artemisia campestris* L. subsp. *variabilis* (Ten.) Greuter are used for post-partum care and as an emmenagogue, analeptic, and antispasmodic. This species is also used as an anthelminthic for stomach and liver afflictions (De Natale & Pollio 2012). *Artemisia herba-alba* Asso is also known as a medicinal plant, whose potions are drunk against digestive troubles; as an antidote to poisons; as an anthelmintic, antiseptic hypoglycemiant, and emmenagogue; and to treat rheumatic pains, afflictions of the eyes, and obesity (Maiza et al. 1996). It is a good fodder as well, but fragments and pollen of plants growing in grazed environments, and particularly Poaceae, were typically more abundant in pastoral sites (Mercuri 2008b; Mercuri et al. 1998).

3.2 *Cereals used for food and fodder*

Poaceae pollen were prominent in all samples. Although this is common in pollen spectra from arid lands, in these sites the high percentages of grass pollen are usually associated with large amounts of well-preserved macroremains. Plant parts, fruits, and pollen of the Paniceae are the most abundant botanical remains in these sites. The common recovery of well-preserved, desiccated remains of *Brachiaria, Cenchrus, Echinochloa, Panicum, Setaria,* and *Urochloa* provides unambiguous evidence that wild cereals were associated with archaeological deposits (Castelletti et al. 1999; Mercuri 2001; Wasylikowa 1992), because they were systematically transported into the shelters (Mercuri 2008a) (Fig. 6).

The early Holocene knowledge, exploitation, selection, care, and processing of wild cereals is clearly recognized from Epipaleolithic to Mesolithic archaeobotanical contexts (e.g. Mercuri & Garcea 2007) for Uan Tabu and Takarkori. Data point to the long-term exploitation of wild cereals and to the prevalence of these plants among those selected and transported to shelters and caves. Human knowledge and uses can be inferred by comparing the set of plants available in the area with the selected samples of plant resources created by humans.

Many species belonging to the genera mentioned above are known to be wild cereals harvested for food and fodder in the Sahara and sub-Sahara. They include *Brachiaria deflexa* (Schumach.) Robyns., *Cenchrus biflorus* Roxb., *Dactyloctenium aegyptium* (L.) Willd., *Latipes senegalensis* Kunth, *Panicum turgidum* Forssk., *P. laetum* Kunth, and other millets, as well as several species of *Eragrostis, Digitaria, Setaria, Andropogon, Bromus,* and *Hyparrhenia* (Harlan 1989). *Panicum turgidum* and *P. repens* L., which currently occur in the Fezzan and the central Sahara (Corti 1942; Ozenda 2000), are useful to bind sand and control soil erosion, and are used as green pasture grasses. *P. turgidum* produces fruits that can be boiled, made into flour or meal, or used for thatch and mats. Powder from the ground stems is used for healing wounds (Williams & Farias 1972). *P. repens* may be cut and harvested five to seven times a year in tropical and subtropical areas, especially to feed domesticated animals (Hossain et al. 2001; Skerman & Riveros 1990).

In the savannah zone, a group of grasses collectively called kreb(s) includes a wild cereal complex—mainly *Panicum laetum* and *Eragrostis pilosa* (L.) P. Beauv.—growing together and therefore producing fruits ripening at the same time (Barakat & Fahmy 1999). They are known to be collected in the Sahel region, stretching from Mali to Sudan (Harlan 1993; Sherif & Siddiqi 1988; Smith 1980).

In North Africa, *Setaria italica* (L.) Beauv. is a minor crop (Zohary & Hopf 2000). *Aristida pungens* L. is the 'drinn', one of the staple cereals, harvested in large amounts to the present day. In West Africa, other millets (*Paspalum scrobiculatum* L. and species of *Pennisetum*) are also harvested.

The genus *Echinochloa* includes a species grown all over the world, namely, *E. crus-galli* (L.) P. Beauv., which presently does not occur either in the Sahara or in Libya, has a good flavour, and is largely used cooked whole or ground into a flour. It also has other uses as medicine or, especially in Egypt, for the reclamation of saline and alkaline areas (Ozenda 2000; Sherif & Siddiqi 1988; Tanaka 1976). *Echinochloa colona* (L.) Link, which is known from the Sahara and Libya, is grown for its fruits, which can be cooked whole or ground into a flour and can also be used mashed or as porridge. The young plants and shoots can be eaten raw or cooked (Manandhar & Manandhar 2002; Tanaka 1976). *Echinochloa stagnina* (Retz.) P. Beauv. is cultivated in the central delta of the Niger River. Its fruits are used as food and to make both alcoholic and non-alcoholic beverages.

In the Acacus archaeological sites wild cereal collection was documented through the recovery of different plant parts, including pollen (flowers), seeds (fruits), and stems and leaves. As mentioned above, high pollen concentrations may be evidence of human transport of these plants. One plant, however, might have been used for several purposes, such as food and basketry.

3.3 Cereals used for basketry

Basketry and cordage made with plant materials were observed in many rock shelter and cave deposits (Cremaschi & di Lernia 1998; Garcea 2001: 92; Maspero 1999). They were in an excellent state of preservation thanks to the limited bacterial activity typical of arid environments, even when perishable artefact material includes fragile items.

In a recent systematic study of basketry from the Takarkori rockshelter, di Lernia et al. (2012) found that the artefacts were made of different organic materials. Stems were generally used for basketry, while a fairly diversified set of materials was used in making ropes and other elements. Panicoideae type *Panicum* and *Setaria* were identified among the prevalent stems used to weave basketry. Therefore, the main plants used for food and fodder have also been employed for weaving.

Interestingly, one piece of basketry was made using a mixture of plant and animal materials, that is, *Panicum* and animal hair. Our informants told us that the women of the Kel Tadrart and Kel Messak, a small lineage of Tuareg herders living today in the Acacus Mountains and Messak Settafet rock desert, make baskets and mats by weaving *Panicum turgidum* and camel hair. The mats are still used during the building of zeribas (typical small huts), especially in sandy desert settlements. The mats are up to four meters long and can take months of work to complete. The camel hairs first have to be twisted or braided to form a sort of thin, strong cordage, and then they are woven with dried stems of *Panicum turgidum*.

4. Conclusions

In the Saharan areas described in this paper, the archaeobotanical record is particularly rich and well preserved. It shows unambiguous evidence of past human handling of plants through the selection and transport of such species as wild cereals, capers, myrtle, sugar date, and ragweed. These very different plants have been specifically preferred for different purposes since the early and middle Holocene. Ethnobotanical observations point to possible uses of these plants in the past, based on their current meaning in popular tradition and on their chemical proprieties. It must be stressed that chemical compounds are intrinsically part of the nature of these plant species and cannot change over time, while cultural knowledge or uses are different for the different cultures.

Therefore, the familiarity with plant records from archaeological contexts suggests not only that ethnobotany (How are plants used today?) is helpful to archaeobotany (Which plants were used in the past?) but also that archaeobotany can make a significant contribution to our knowledge about the continuity and historical depth of ethnobotanical knowledge in a region (How far back do we have to go to find knowledge of these plants?).

Indigenous knowledge systems of pharmacopoeia and other uses of plants both for humans and animals are still of great importance today. Modern ethnobotanical research preserves biocultural knowledge and investigates the potential of the biological diversity of our world (Choudhary et al. 2008). Archaeobotany plays a role in classical ethnobotanical studies by contributing, on the one hand, to research in economic botany and the exploration of past human cognition and, on the other hand, to biodiversity preservation and current vegetation management practices.

Acknowledgements

We give our best thanks to Elena A.A. Garcea (University of Cassino, Italy), for giving us the opportunity to study the palynology of Gobero, and to, initially, Fabrizio Mori; then Mario Liverani; and, subsequently, Savino di Lernia of Sapienza University of Rome and the Department of Archaeology, Tripoli, who is among the directors of the Italian–Libyan Archaeological Mission in the Acacus and Messak, for entrusting to AMM the botanical material of the mission.

We are grateful to the Tuareg people for their invaluable help. Many thanks to Ursula Thanheiser for organizing the 7th IWAA and putting together this book. We are truly indebted to Catherine D'Andrea for her great help and for improving the English; her comments, together with those of another, anonymous referee, greatly improved the clarity of the text.

References

Adam J.G., Echard N. & Lescot M., 1972. Plantes médicinales Hausa de l'Ader (République du Niger). *Journal d'Agriculture Tropicale et de Botanique Appliquée* 19, pp. 259-399.

Adjanohoun E., Ahyi M.R.A., Ake Assi L., Baniakina J., Chibon P., Cusset G., Doulou V., Enzanza A., Eymé J., Goudoté E., Keita A., Mbemba C., Mollet J., Moutsambote J.M., Mpati J. & Sita P., 1988. *Contribution aux études ethnobotaniques et floristiques en République populaire du Congo.* Agence de coopération culturelle et technique, Paris.

Bandini Mazzanti M., Bosi G., Mercuri A.M., 2005. Paleoeconomia vegetale: Origine dell'agricoltura e delle specie coltivate. In: Caneva G. (ed.), *La biologia vegetale per i beni culturali II, Conoscenza e valorizzazione.* Nardini Editore, Firenze, pp. 426-428.

Barakat H.N. & Fahmy A.G., 1999. Wild grasses as 'Neolithic' food resources in the eastern Sahara: A review of the evidence from Egypt. In: Van der Veen M. (ed.), *The Exploitation of Plant Resources in Ancient Africa.* Kluwer, New York, pp. 33-46.

Bekalo T.H., Woodmatas D.S. & Woldemariam A.Z., 2009. An ethnobotanical study of medicinal plants used by local people in the lowlands of Konta Special Woreda, southern nations, nationalities and peoples regional state, Ethiopia. *Journal of Ethnobiology and Ethnomedicine* 5, pp. 26.

Bellakhdar J., Claisse R., Fleurentin J. & Younos C., 1991. Repertory of standard herbal drugs in the Moroccan pharmacopoea. *Journal of Ethnopharmacology* 35, pp.123-143.

Bernus E., 1992/93. Des arbres et des herbes aux marges du Sahara. *Sahara* 5, pp. 17-28.

Bora S.K. & Sharma A., 2011. The genus *Artemisia*: A comprehensive review. *Pharmaceutical Biology* 49, pp. 101-109.

Bottema S., 1992. Prehistoric cereal gathering and farming in the Near East: The pollen evidence. *Review of Palaeobotany and Palynology* 73, pp. 21-33.

Boulos L., 1983. *Medicinal Plants of North Africa.* Medicinal Plants of the World 3. Reference Publications, Algonac, Michigan.

Byavu N., Henrard C., Dubois M. & Malaisse F., 2000. Phytothérapie traditionnelle des bovins dans les élevages de la plaine de la Ruzizi. *Biotechnology Agronomy Society and Environment* 4, pp. 135-156.

Cappers R.T.J., 2006. *Roman Foodprints at Berenike: Archaeobotanical Evidence of Subsistence and Trade in the Eastern Desert of Egypt.* Cotsen Institute of Archaeology Monograph 55, University of California, Los Angeles.

Cappers R.T.J., Sikking L., Darnell J.C. & Darnell D., 2007. Food supply along the Theban desert roads (Egypt): The Gebel Roma', Wadi el-Hôl, and Gebel Qarn el-Gir caravansary deposits. In: Cappers R.T.J. (ed.) Fields of Change: Progress in African Archaeobotany. [Proceedings of the 4th International Workshop on African Archaeobotany.] Groningen Archaeological Studies 5, Barkhuis & Groningen University Library, Groningen, pp. 127-138.

Cappers R.T.J. & Neef R., 2012. *Handbook of Plant Palaeoecology.* Barkhuis & Groningen University Library, Groningen.

Castelletti L., Castiglioni E., Cottini M. & Rottoli M., 1999. Archaeobotanical analysis of charcoal, wood and seeds. In: di Lernia S. (ed.), *The Uan Afuda Cave Hunter-Gatherer Societies of Central Sahara.* All'Insegna del Giglio, Firenze, pp. 131-148.

Cattabiani A., 1998. *Florario: Miti, leggende e piante.* Mondadori, Milano.

Corti R., 1942. *Flora e vegetazione del Fezzan e della regione del Gat.* Ricci, Firenze.

Cremaschi M. & di Lernia S., 1998. *Wadi Teshuinat Palaeoenvironment and Prehistory in South-Western Fezzan (Libyan Sahara).* Quaderni Geodinamica Alpina e Quaternaria. C.N.R., Firenze.

Cremaschi M. & Zerboni A., 2009. Early to middle Holocene landscape exploitation in a drying environment: Two case studies compared from the central Sahara (SW Fezzan, Libya). *Comptes Rendus Géoscience* 341, pp. 689-702.

Cremaschi M. & Zerboni A., 2010. Human communities in a drying landscape: Holocene climate change and cultural response in the central Sahara. In: Martini I.P. & Chesworth W. (eds.), *Landscapes and Societies.* Springer, Dordrecht, pp. 67-89.

Cunningham A.B., 1997. An Africa-wide overview of medicinal plant harvesting, conservation and health care. In: Bodeker G., Bhat K.K.S., Burley J. & Vantomme P. (eds.), *Medicinal Plants for Forest Conservation and Health Care.* Non-Wood Forest Products 11. Daya, Rome, pp. 116-129.

De Natale A. & Pollio A., 2012. A forgotten collection: The Libyan ethnobotanical exhibits (1912–14) by A. Trotter at the Museum O. Comes at the University Federico II in Naples, Italy. *Journal of Ethnobiology and Ethnomedicine* 8 (4). doi: 10.1186/1746-4269-8-4

Diamond J., 2002. Evolution, consequences and future of plant and animal domestication. *Nature* 418, pp. 700-707.

di Lernia S., 1999. *The Uan Afuda Cave: Hunter-Gatherer Societies of Central Sahara.* All'Insegna del Giglio, Firenze.

di Lernia S., 2001. Dismantling dung: Delayed use of food resources among early Holocene foragers of the Libyan Sahara. *Journal of Anthropological Archaeology* 20, pp. 408-441.

di Lernia S., 2002. Dry climatic events and cultural trajectories: Adjusting middle Holocene pastoral economy of the Libyan Sahara. In: Hassan F. (ed.), *Droughts, Food and Culture: Ecological Change and Food Security in Africa's Later Prehistory.* Kluwer, New York, pp. 225-250.

di Lernia S., Massamba N'siala I. & Zerboni A., 2012. "Saharan waterscapes": Traditional knowledge and historical depth of water management in the Akakus Mts. (SW Libya). In: Sternberg T. & Mol L. (eds.), *Changing Deserts: Integrating People and Their Environment.* Cambridge, White Horse Press, pp. 101-128.

di Lernia S. & Zampetti D., 2008. *La memoria dell'arte: Le pitture rupestri dell'Acacus tra passato e futuro.* All'Insegna del Giglio, Firenze.

Erasto P., Adebola P.O., Grierson D.S. & Afolayan A.J., 2005. An ethnobotanical study of plants used for the treatment of diabetes in the Eastern Cape Province, South Africa. *African Journal of Biotechnology* 4, pp. 1458-1460.

Faegri K. & Iversen J., 1989. *Textbook of Pollen Analysis.* 4th ed. Wiley, Chichester.

Garba H., 2000. Les produits forestiers non-ligneux au Niger : connaissances actuelles et tendances. FAO, online http://www.fao.org/DOCREP/003/X6740F/X6740F00.HTM

Garcea E.A.A., 2001. The Pleistocene and Holocene archaeological sequence. In: Garcea E.A.A. (ed.), *The Uan Tabu in the Settlement History of the Libyan Sahara.* Edizioni all'Insegna del Giglio, Firenze, pp. 25-49.

Garcea E.A.A., 2013. *Gobero: The No-Return Frontier: Archaeology and Landscape at the Saharo-Sahelian Borderland.* Journal of African Archaeology Monograph Series 9. Africa Magna Verlag, Frankfurt.

Gasse F., 2000. Hydrological changes in the African tropics since the Last Glacial Maximum. *Quaternary Science Reviews* 19, pp. 189-211.

Giraudi C. & Mercuri A.M., 2013. Early to middle Holocene environmental variations in the Gobero basin. In: Garcea E.A.A. (ed.), *Gobero: The No-Return Frontier: Archaeology and Landscape at the Saharo-Sahelian Borderland.* Journal of African Archaeology Monograph Series 9, Africa Magna Verlag, Frankfurt, pp. 114-126.

Giraudi C., Mercuri A.M. & Esu D., 2013. Holocene palaeoclimate in the northern Sahara margin (Jefara Plain, northwestern Libya). *The Holocene* 1, pp. 339-352.

Gour V.S. & Kant T., 2012. *Balanites aegyptiaca* (L.) Del.: A multipurpose and potential biodiesel tree species of the arid regions. *International Journal of Science and Nature* 3, pp. 472-475.

Hamidou T.H., Kabore H., Ouattara O., Ouédraogo S., Guissou I.P. & Sawadogo L., 2002. Efficacy of *Balanites aegyptiaca* (L.) Del. Balanitaceae as anthelminthic and molluscicid used by traditional healers in Burkina Faso. Paper presented at the International Conference on Emerging Infectious Diseases, Atlanta [abstract p. 37].

Hammiche V. & Maiza K., 2006. Traditional medicine in central Sahara: Pharmacopoeia of Tassili N'ajjer. *Journal of Ethnopharmacology* 105, pp. 358-367.

Harlan J.R., 1989. Wild-grass seeds harvesting in the Sahara and sub-Sahara of Africa. In: Harris D. & Hillman G.C. (eds.), *Foraging and Farming, the Evolution of Plant Exploitation.* Unwin Hyman, London, pp. 79-88.

Harlan J.R., 1993. Genetic resources in Africa. In: Janick J. & Simon J.E. (eds.), *New Crops.* Wiley, New York, pp. 65-67.

Harouna H., Favel A., Delmas F., Elias R., Saadou M. & Balansard G., 1996. Screening of some plants used in traditional medicine in Niger for antiparasitic and antifungal activities. In: Schröder E., Balansard G., Cabalion P., Fleurentin J. & Mazars G. (eds.), *Medicaments et aliments—Approche ethnopharmacologique.* [Actes du 2e Colloque Européen d'ethnopharmacologie et de la 11e Conférence internationale d'ethnomédecine, Heidelberg]. Orstom Éditions, Paris, pp. 285-287.

Head L.M., Atchison J.M. & Fullagar R.L., 2002. Country and garden: Ethnobotany, archaeobotany and Aboriginal landscapes near the Keep River, northwestern Australia. *Journal of Social Archaeology* 2, pp. 173-196.

Hoelzmann P., Schwald A., Roberts N., Cooper P. & Burgess A., 2010. Hydrological response of an east-Saharan palaeolake (NW Sudan) to early-Holocene climate. *The Holocene* 20, pp. 537-549.

Hossain M.A., Akamine H., Nakamura I., Ishimine Y. & Kuramochi H., 2001. Influence of temperature levels and planting time on the sprouting of rhizome-bud and biomass production of torpedo grass (*Panicum repens* L.) in Okinawa Island, southern Japan. *Weed Biology and Management* 1, pp. 164-169.

Iwu, M.M., 1993. *Handbook of African Medicinal Plants.* CRC Press, Boca Raton.

Jaouad E.-H., Hmammouchi M. & Lyoussi B., 2003. Ethnobotanical studies and economic evaluation of medicinal plants in Taounate province (northern Morocco). *Journal of Ethnopharmacology* 86, pp. 149-158.

Kröpelin S., Verschuren D. & Lézine A.M., 2008. Climate-driven ecosystem succession in the Sahara: The past 6000 years. *Science* 320, pp. 765-768.

Lézine A.M., 2009. Climatic history of the African and Arabian deserts. *Comptes Rendus Géoscience* 341, pp. 569-574.

Lézine A.M., Hély C., Grenier C., Braconnot P. & Krinner G., 2011. Sahara and Sahel vulnerability to climate changes, lessons from Holocene hydrological data. *Quaternary Science Review* 30, pp. 3001-3012.

Luxereau A., 1994. Usages, representations, évolution de la biodiversité végétale chez les Haoussa du Niger. *Journal d'Agriculture Traditionalle et de Botanique Appliquée* 36, pp. 67-85.

Maiza K., Brac De La Perrière R.A. & Hammiche V., 1996. Pharmacopée traditionnelle saharienne: Sahara septentrional. In: Schröder E., Balansard G., Cabalion P., Fleurentin J. & Mazars G. (eds.), *Medicaments et aliments—approche ethnopharmacologique.* [Actes du 2e Colloque Européen d'ethnopharmacologie et de la 11e Conférence internationale d'ethnomédecine, Heidelberg.] Orstom Éditions, Paris, pp. 169-171.

Maley J., 1972. La sedimentation pollinique actuelle dans le zone du lac Tchad. *Pollen et Spores* 14, pp. 263-307.

Manandhar N.P. & Manandhar S., 2002. *Plants and People of Nepal.* Timber Press, Portland.

Martin G.J., 1995. *Ethnobotany: A Methods Manual.* Chapman & Hall, London.

Maspero A., 1999. Spinning and plaiting. In: di Lernia S. (ed.), *The Uan Afuda Cave: Hunter-Gatherer Societies of Central Sahara.* Arid Zone Archaeology Monographs 1. All'Insegna del Giglio, Firenze, pp. 183-87, 239-253.

McGovern P.E., Christofidou-Solomidou M., Wang W., Dukes F., Davidson T. & El-Deyri W.S., 2010. Anticancer activity of botanical compounds in ancient fermented beverages (Review). *International Journal of Oncology* 37, pp. 5-14.

Mercuri A.M., 1999. Palynological analysis of the Early Holocene sequence. In: di Lernia, S. (ed.), *The Uan Afuda Cave: Hunter-Gatherer Societes of Central Sahara.* Arid Zone Archaeology Monographs 1. All'Insegna del Giglio, Firenze, pp. 149-181.

Mercuri A.M., 2001. Preliminary analyses of fruits, seeds and few plant macrofossils from the Early Holocene sequence. In: Garcea, E.A.A. (ed.), *Uan Tabu in the Settlement History of the Libyan Sahara.* Arid Zone Archaeology Monographs 2. All'Insegna del Giglio, Firenze, pp. 189-210.

Mercuri A.M., 2008a. Plants and culture: A neglected basic partnership for interculturality. In: Morel J.-P. & Mercuri A.M. (eds.), *Plants and Culture: Seeds of the Cultural Heritage of Europe.* Centro Europeo per i Beni Culturali, Ravello/Edipuglia, Bari, pp. 17-24.

Mercuri A.M., 2008b. Plant exploitation and ethnopalynological evidence from the Wadi Teshuinat area (Tadrart Acacus, Libyan Sahara). *Journal of Archaeological Science* 35, pp. 1619-1642.

Mercuri A.M., 2008c. Human influence, plant landscape, evolution and climate inferences from the archaeobotanical records of the Wadi Teshuinat area (Libyan Sahara). *Journal of Arid Environment* 72, pp. 1950-1967.

Mercuri A.M. & Garcea E.A.A., 2007. The impact of hunter/gatherers in the central Sahara during the early Holocene. In: Cappers R.T.J. (ed.) *Fields of Change. Progress in African Archaeobotany.* [Proceedings of the 4th International Workshop on African Archaeobotany.] Groningen Archaeological Studies 5, Barkhuis & Groningen University Library, Groningen, pp. 87-104.

Mercuri A.M., Massamba N'siala I. & Florenzano A., 2013. Environmental and ethnobotanical data inferred from pollen of Gobero and the dried lakebeds in the surrounding area. In: Garcea E.A.A. (ed.), *Gobero: The No-Return Frontier: Archaeology and Landscape at the Saharo-Sahelian Borderland.* Journal of African Archaeology Monograph Series 9, Africa Magna Verlag, Frankfurt, pp. 81-104.

Mercuri A.M. & Sadori L., 2014. Mediterranean culture and climatic change: Past patterns and future trends. In: Goffredo S. & Dubinsky Z. (eds.), *The Mediterranean Sea: Its History and Present Challenges.* Springer, Dordrecht, pp. 507-527.

Mercuri A.M., Sadori L. & Uzquiamo Ollero P., 2011. Mediterranean and north-African cultural adaptations to mid-Holocene environmental and climatic changes. *The Holocene* 21 (1), pp. 189-206.

Mercuri A.M., Trevisan Grandi G., Mariotti Lippi M. & Cremaschi M., 1998. New pollen data from the Uan Muhuggiag rockshelter (Libyan Sahara, VII–IV millennia BP). In: Cremaschi M. & di Lernia S. (eds.), *Wadi Teshuinat: Palaeoenvironment and Prehistory in South-Western Fezzan (Libyan Sahara).* CNR–Quaderni di Geodinamica Alpina e Quaternaria, Milano, pp. 107-124.

Mercuri A.M., Trevisan Grandi G., Bosi G., Forlani L. & Buldrini F., 2006. The archaeobotanical remains pollen, seeds/fruits and charcoals. In: Liverani M. (ed.), *The Archaeology of Libyan Sahara.* Volume II—Aghram Nadharif: The Barkat Oasis (Sha'abiya of Ghat, Libyan Sahara) in Garamantian Times. All'Insegna del Giglio, Firenze, pp. 335-348.

Merzouki A., Ed-derfoufi F. & Molero Mesa J., 2000. Contribution to the knowledge of Rifian traditional medicine: II: Folk medicine in Ksar Lakbir district (NW Morocco). *Fitoterapia* 71 (3), pp. 278-307.

Minnis P.E., 2000. *Ethnobotany: A Reader.* University of Oklahoma Press, Norman.

Mishra S.N., Tomar P.C. & Lakra N., 2007. Medicinal and food value of *Capparis*—a harsh terrain plant. *Indian Journal of Traditional Knowledge* 6, pp. 230-238.

Ndoye M., Diallo I. & Gassama/Dia Y.K., 2004. Reproductive biology in *Balanites aegyptiaca* (L.) Del., a semi-arid forest tree. *African Journal of Biotechnology* 3, pp. 40-46.

Neuwinger H.D., 2000. *African Traditional Medicine.* Medpharm Scientific Publishers, Stuttgart.

Olmi L., Massamba N'siala I., Buldrini F., Mariotti Lippi M., di Lernia S. & Mercuri A.M., 2007. Cereali selvatici a Takarkori, sito del Tadrart Acacus—Sahara libico: Schede di alcuni macroresti da uno spot dell'Olocene Medio. *Atti Società Naturalisti e Matematici di Modena* 137, pp. 411-430.

Olmi L., Mercuri A.M., Gilbert M.T.P., Biagetti S., Fordyce S., Cappellini E., Massamba N'siala I. & di Lernia S., 2011. Morphological and genetic analyses of early mid-Holocene wild cereals from the Takarkori rockshelter (central Sahara, Libya)—first results and prospects. In: Fahmy A.G. & D'Andrea C. (eds.), *Windows on the African Past: Current Approaches to African Archaeobotany.* [Proceedings of the sixth International Workshop for African Archaeobotany.] Africa Magna Verlag, Frankfurt, pp. 175-184.

Olmi L., Sala S., Arru L. & Mercuri A.M., 2012 . Uniformità morfobiometrica in accumuli di cereali selvatici dell'Olocene iniziale nel Sahara centrale: Spighette di *Panicum, Echinochloa* e *Sorghum.* [Programma e Riassunti VII Congresso Nazionale di Archeometria A.I.Ar. 22–24 Febbraio 2012. Università di Modena e Reggio Emilia Modena ITA.] p. 58.

Okigbo R.N., Eme U.E. & Ogbogu S., 2008. Biodiversity and conservation of medicinal aromatic plants in Africa. *Biotechnology and Molecular Biology Reviews* 3, pp. 127-134.

Ozenda P., 2000. *Flore et végétation du Sahara.* CNRS, Paris.

Pearsall D.M., 2000. *Paleoethnobotany: A Handbook of Procedures.* 2nd ed. Academic Press, New York.

Prance G.T., 1991. "What is ethnobotany today?" *Journal of Ethnopharmacology* 32, pp. 209-216.

Schultes R.E. & von Reis S., 1995. *Ethnobotany: Evolution of a Discipline.* Timber Press, Portland.

Sereno P.C., Garcea E.A.A., Jousse H., Stojanowski C.M., Saliège J.F., Maga A., Ide O.A., Knudson K.J., Mercuri A.M., Stafford T.W., Kaye, Jr., T.G., Giraudi C., Massamba N'siala I., Cocca E., Moots H.M., Dutheil D.B. & Stivers J.P., 2008. Lakeside cemeteries in the Sahara: 5000 years of Holocene population and environmental change. *PLoS ONE* 3 (8), e2995.

Sherif A.S. & Siddiqi M.A., 1988. Poaceae. In: El-Gadi A.A. (ed.), *Flora of Libya.* Al Faateth University, Tripoli.

Skerman P.J. & Riveros F., 1990. *Tropical Grasses.* FAO Plant Production and Protection Series 23. FAO, Rome.

Smith S.E., 1980. The environmental adaptation of nomads in the West African Sahel: A key to understanding prehistoric pastoralists. In: Williams M.A.J. & Faure H., *The Sahara and the Nile: Quaternary Environments and Prehistoric Occupation in Northern Africa.* A.A. Balkema, Rotterdam, pp. 467-487.

Société française d'ethnopharmacologie, 1990. www.ethnopharmacologia.org.

Tanaka T., 1976. *Tanaka's cyclopedia of edible plants of the world.* Keigaku Publishing Company, Tokyo.

Van der Veen M., 1995. Ancient agriculture in Libya: A review of the evidence. *Acta Palaeobotanica* 35, pp. 85-98.

Van der Veen M. (ed.) 1999. *The Exploitation of Plant Resources in Ancient Africa.* New York, Kluwer Academic/Plenum Publishers.

Van der Veen M., 2007. Formation processes of desiccated and carbonized plant remains—the identification of routine practice. *Journal of Archaeological Science* 34, pp. 968-990.

Van der Veen M., 2011. Consumption, trade and innovation: Exploring the botanical remains from the Roman and Islamic Ports at Quseir al-Qadim, Egypt. Journal of African Archaeology Monograph Series 6. Africa Magna Verlag, Frankfurt.

Wasylikowa K., 1992. Holocene flora of the Tadrart Acacus area, SW Libya, based on plant macrofossils from Uan Muhuggiag and Ti-n-Torha Two Caves archaeological sites. *Origini* 16, pp. 125-159.

White F., 1983. *The Vegetation of Africa.* UNESCO, Paris.

Williams J.T. & Farias R.M., 1972. Utilisation and taxonomy of the desert grass *Panicum turgidum. Economic Botany* 26, pp.13-20.

Zohary D. & Hopf M., 2000. *Domestication of Plants in the Old World.* 3rd ed. University Press, New York & Oxford.

Taro across the oceans: Journeys of one of our oldest crops

Ilaria Maria Grimaldi

Research Laboratory for Archaeology and the History of Art
University of Oxford
New Barnett House
28 Little Clarendon Street
Oxford OX1 2HU, United Kingdom

E-mail: marinastella50@hotmail.com

Abstract. Taro is an ancient crop native to Asia that is now found in all tropical and warm temperate regions of the world. Its present distribution is the result of a long history of use and of transoceanic translocation. While in the Pacific region much effort has been made to understand the role this plant played in prehistory, in Africa it has not received the same attention, even though it is a staple crop in some West African countries. Recently, the retrieval of macroremains of desiccated taro at the Islamic port of Quseir al-Qadim in Egypt (Van der Veen & Morales 2011) has revitalised the debate on the history of taro in Africa, and genetic research has been initiated to reconstruct its introduction into the continent. Fieldwork and a genetic survey of taro in Africa were carried out as part of a project led by Nicole Boivin (University of Oxford) and Robin G. Allaby (University of Warwick). Although analysis of the genetic survey results is ongoing, a geographical and cultural appraisal of taro has now been completed and is presented here.

Keywords: taro, crop dispersal, ethnobotany, Africa

1. Introduction

Cultivated for its edible starchy rhizomes, taro[1] has long been neglected by research and considered a minor crop, despite probably being one of the oldest cultivated crops (Lebot 2009; Porterfield 1944). In the Solomon Islands, excavations at the site of Kilu Cave, dated to between 28 000 and 20 000 BP, have uncovered prehistoric stone tools on which likely taro starch grains and calcium oxalate crystals were observed (Loy et al. 1992). Starch grains are also reported from Kuk Swamp, in Papua New Guinea (Fullagar et al. 2006), supporting the view that taro has a long history of use and may have played an important role in the early stages of agricultural development. The natural geographical range of wild taro appears to stretch from north-eastern India to Australia and Papua New Guinea (Burkill 1935; Matthews 1991). Whether humans used and transported wild forms of taro remains an open question. Large expanses of ocean may have restricted a natural dispersal (Matthews et al. 1992), but island bridges in the South-East Asia region may have facilitated natural dispersal of wild progenitors, creating opportunities for independent domestication events in different regions (Denham 2010; Hotta 2002; Matthews 1990; Yoshino 2002).

In the Pacific region, the strong association between Austronesian-speaking people and the dispersal of taro across the Pacific (e.g. Bellwood 2006; Matthews 1995) suggests that they may have domesticated the plant somewhere in South-East Asia. Subsequently, taro may

1 Generally speaking, taro is an English term adopted from Polynesian languages and applied loosely to other species in the same plant family (*Alocasia macrorrhizos* (L.) G.Don, *Cyrthosperma merkusii* (Hassk.) Schott, and *Xanthosoma sagittifolium* (L.) Schott). In this paper, the term refers exclusively to *Colocasia esculenta* (L.) Schott.

have been introduced into Africa by more than one route and at different times. The local belief in West Africa is that it is an indigenous crop (Dalziel 1937; Williamson 1970; author's field notes). In Egypt (de Sacy 1810), Palestine, Syria (Cook 1910), and Cyprus (Anguillara 1561; Matthews 2005), taro was already being cultivated many centuries ago, but it has never been found in depictions on early Egyptian monuments, suggesting that this plant was not present during pharaonic times (Täckholm & Drar 1941).

In the literature, the earliest reference to taro may be from the second century BC in a Chinese book, where the plant was mentioned as an emergency crop used during times of famine (Huang 2012). Its presence in the Mediterranean has never been reviewed. It was long thought that the plant named *colocasia* (Greek: κολοκασιον) in Greek and Latin texts (Dioscorides; Pliny) was the modern plant *Colocasia esculenta*. However, descriptions by ancient scholars (Dioscorides MM 2,128; Pliny NH 21,87) seem to indicate, rather, that the name was initially applied to the rhizomes of the sacred Indian lotus plant, *Nelumbo nucifera* Gaertn., known at the time as the 'Egyptian bean', for its edible seeds that are similar to beans and that were highly praised among the Alexandrians (Nic *apud* Ath. 3,72a). After the fourth century AD, the Latin name *colocasia* seems to have been applied to taro (Burkill 1938). Palladius referred to *colocasia* bulbs as being planted in February (Pall. Op. Agr. 3,24,14), and the cultivation recommendations suggest that this was taro rather than the lotus plant.

We know that taro was certainly present in Egypt from the second half of the first millennium AD onward, based on both textual (Al-Masudi 1864; Silvatico 1474) and archaeological evidence (Van der Veen & Morales 2011). The recent discovery of macroremains of this plant at the archaeological site of Quseir al-Qadim attests to its presence during the Islamic occupation, with radio-carbon dates confirming its earliest presence at AD 1050–1170 (Van der Veen & Hamilton 2011: Appendix 1; Van der Veen & Morales 2011: 95). By the fourteenth century AD taro had reached West Africa, from where it was eventually introduced to the Americas, travelling across the Atlantic Ocean in the seventeenth century during the slave trade, as provision for the journey (Carney & Rosomoff 2009).

Human networks in which taro was transported spanned the Indian, Pacific and Atlantic oceans, as well as the Mediterranean Sea, over a period of thousands of years. This paper explores just a few cultural and historical aspects of the relationship between taro and humans and aims to reconstruct its travels around the globe.

2. Morphology, uses, and taxonomy

Taro, *Colocasia esculenta* (L.) Schott, belongs to the family Araceae, also known as the aroids, or *Arum* family. The family is globally distributed from temperate to tropical regions. The inflorescence is composed of a spadix surrounded by a large bract, called spathe. The spadix is composed of an upper part with sessile male flowers and a lower part with sessile female flowers. When pollination occurs, the plant produces fruiting heads with many berries containing numerous small seeds (Plucknett 1970). Seed production in taro in cultivation is rare, and as yet none has been reported archaeologically. Taro leaves are heart-shaped and can grow to a substantial size, with blades 0.5–1 m in length and petioles up to 2 m in height. Although the leaves show variable morphology, they all have a peltate structure, with the petiole attached centrally to the lower surface of the leaf. This is a useful characteristic for distinguishing *C. esculenta* from similar aroids, particularly *Alocasia macrorrhizos* (L.) G.Don and *Xanthosoma sagittifolium* (L.) Schott, both of which are often misidentified as *C. esculenta*. The petioles of these other aroids are attached to the edge of a sagittate leaf blade.

The main value of taro as a crop today is in the presence of starchy corms (rhizomes). In some cultures, the blades and petioles are eaten, but it is mainly the corm that is used for culinary purposes and that has commercial value. Purseglove (1972) analysed the morphology of the corms and distinguished two main forms. The first is called dasheen, a term said to be derived from the French *Taro de Chine* (Taro of China), which eventually became shortened to de Chine and subsequently to dasheen (Young 1917). It is characterized by having a large, edible corm and it is better adapted to flooded swampland environments (Lebot et al. 2004). The second form, known as eddoe, produces a small to medium-sized main corm (which is often discarded because of its acridity) and a large number of small, edible cormels (Plucknett 1983). This type grows best in rain-fed conditions and does not adapt well to swampy environments (Lebot et al. 2004). However, the local terminology changes according to geographical area, and many intermediate forms exist that cannot be assigned to either of these two groupings. Further genetic research may shed light on this issue.

To some extent, taxonomic complications linked to the scientific name *Colocasia esculenta* reflect the polymorphic nature of taro. In *Species Plantarum*, Linnaeus (1753) described two types of taro: *Arum Colocasia*, from specimens growing on the Mediterranean rim, and *Arum esculentum*, from specimens collected in America. The specific name *Colocasia* used by Linnaeus comes from the belief that taro was the *colocasia* mentioned by Greek and Latin authors in the classical literature, whereas the specific name *esculentum* means edible. Later, Schott (1832) established a new genus that he named *Colocasia*, to which he ascribed the names *Arum Colocasia*, *Arum esculentum*, and *Caladium acre* R. Br., which he eventually renamed *Colocasia antiquorum*, *Colocasia esculenta*, and *Colocasia acris*. However, the question marks he inserted before the names *C. esculenta* and *C. acris* seem to express his doubt that these specimens belonged to different species. As a matter of fact, in 1856, Schott reassigned these species to the polymorphic *Colocasia antiquorum* and reduced the others to the taxonomic rank of variety. However, as Hill pointed out (1939), according to the nomenclature rules, the use of the name *Colocasia antiquorum* is incorrect, as it is not possible to reduce the older specific name *esculentum* to a varietal rank, and that the ranks should be reversed and *Colocasia esculenta* should be used instead.

Apart from several formal taxonomic varieties (International Plant Names Index), the current nomenclature does not recognise subdivisions within the species *Colocasia esculenta*, and the often-mentioned *C. esculenta* var. *antiquorum* and *C. esculenta* var. *esculenta* are not currently considered distinct. Until quite recently, the genus *Colocasia* contained 11 recognised species (Govaerts & Frodin 2002). Further species have been reported by botanists in China (Cao & Long 2003; Yin 2006; Yin et al. 2004), and today, The Plant List (2013), a database compiled by the Royal Botanic Gardens at Kew and Missouri Botanical Garden, recognises 16 accepted species names within the genus *Colocasia*. It is likely that many new or newly reported species will gain recognition as fieldwork and laboratory research continue.

3. Genetic perspective and origins

From a genetic point of view, research on taro has been neglected compared with other staple crops. Nevertheless, in recent decades, genetic research on aroids has gained momentum (Ahmed et al. 2012; Caillon et al. 2006; Irwin et al. 1998; Kreike et al. 2004; Lebot & Aradhya 1991; Lebot et al. 2004; Mace & Godwin 2002; Mace, Mathur, Godwin et al. 2006; Mace, Mathur, Izquierdo et al. 2006; Matthews & Terauchi 1994; Nguyen et al. 1998; Singh et al. 2008).

Karyological investigations (Coates et al. 1988; Kuruvilla & Singh 1981; Matthews 1990; Parvin et al. 2008; Yen & Wheeler 1968) have shown that taro includes two levels of polyploidy, diploid (2n=28) and triploid (2n=42). Other aneuploid or polyploid numbers are rarely reported (2n=21,22,24,26,29,38,56). Triploids are associated with numerous small cormels, whereas diploids are generally composed of a single main corm (Okonkwo 1993). Diploids are better adapted to warm and humid environments (Yoshino 2002) and are more common in Asia and Oceania (Matsuda & Nawata 2002). On the other hand, triploids, generally thought to be of Asian origin (Matthews 1990; Yen & Wheeler 1968), seem to be better adapted to colder and drier environments (Yoshino 2002) and are generally found in the higher altitudes or latitudes stretching from Nepal to Japan (Matsuda & Nawata 2002). In triploids, the mechanism of genome duplication is believed to have arisen through the production of gametes that have not had their ploidy level reduced during meiosis (Matthews 1990; Yoshino 2002). Even so, other mechanisms of polyploidy have been detected, such as the production of triploids from self-pollinated populations in which both pollen parent and seed parent can serve as genome donor (Isshiki et al. 1999).

Polyploidy is common in most crops (e.g. Léon 1977), and this trait appears to have conferred ecological success on cultivated plants (Hilu 1993). The question of whether cultivated taro was domesticated once or multiple times is the subject of ongoing research. In this regard the common tropical wildtype *Colocasia esculenta* var. *aquatilis* (Hassk.) has been put forward as a possible progenitor of the widely cultivated *Colocasia esculenta* (L.) Schott (Barrau 1965; Hotta 1970; Matthews 1991). This plant is characterised by small, acrid, white corms and long running stolons that assure its vegetative dispersal and persistence in undisturbed environments. It grows alongside streams and waterfalls, and it flowers in sunny and warm environments (Matthews 1998). While it is less commonly consumed by humans, the leaves, once cooked, can be mixed with rice to be used as pig fodder (Matthews 2005).

Wild taro populations have been found in India, Nepal, southern China, Burma, South-East Asia, the Okinawa islands, New Guinea, and northern Australia (Matthews 1991, 1995, 2005; Matthews et al. 1992), suggesting a wide natural area of distribution in which domestication could have repeatedly occurred. However, the presence of the wildtype taro in the Ryukyu archipelago in Japan raises the question of whether the plant was transported by humans across the ocean (Matthews 1998). Recent linguistic research on the vernacular names applied to taro in the Indo-Pacific area shows three linguistic centres, which could represent many domestication centres (Blench 2012). Thus far, distinctive taro gene pools have been identified in Nepal (Nguyen et al. 1998; Yoshino 2002), in the South-East Asia region (Lebot et al.

2004), and in the Pacific area (Mace, Mathur, Godwin et al. 2006; Mace, Mathur, Izquierdo et al. 2006), with some overlap in Indonesia (Irwin et al. 1998; Kreike et al. 2004; Lebot & Aradhya 1991). These genetic data seem to support the results of the linguistic research.

Purseglove (1972) noted that Pacific cultivations, when compared with South-East Asian cultivations, appear more homogeneous and that only one type of taro (the so-called *C. esculenta* var. *esculenta*) was present. Lebot et al. (2004) noted that eddoe types (with cormels) are not found in Melanesia, probably because of their lack of environmental adaptation. In a recent ethnobotanical study of taro in Ethiopia, Fujimoto (2009), citing Safo Kantaka (2004), noted that in Africa taro is represented mostly by the eddoe type, probably for its better adaptation to drier conditions.

In the Pacific region, production and genetic diversity of taro has recently been undermined by catastrophic outbreaks of taro leaf blight—a disease caused by the fungus (oomycete) *Phytophthora colocasiae*. Less resistant cultivars have been lost in some areas (for a complete review on taro leaf blight, see Singh et al. [2012]). Breeding and conservation programs have been established to restore taro production and preserve genetic diversity of taro in the area (Taylor et al. 2010). Recently, taro leaf blight has also been found in West Africa (author's observation; Bandyopadhyay et al. 2011; Guarino 2010; Omane et al. 2012). Disappearing cultivars are now likely to be replaced by plants of yautia, a New World aroid that is usually preferred for its taste, ease of cooking, and capacity to adapt to savannah-type environments (author's fieldwork), and by taro plants of Pacific provenance (personal communication, Leke Nkeabeng, The Institute of Agricultural Research for Development, IRAD-Cameroon, February 2011).

A collection of taro leaf samples gathered in 2010, during the author's African fieldwork, now represents an important snapshot of genetic diversity of taro in Africa before the outbreak of *Phytophthora*. Future biogeographical studies will need to consider the reshuffling caused by the disappearance and replacement of cultivars.

4. Taro in the Pacific

In the Pacific region, taro is one of the most historically and spiritually significant crops. In many areas, the use of terraces and irrigation systems to grow taro shows that the plant occupies a significant place in the social order (e.g. Sand 2012; Spriggs 2012). The significance of this plant to a society's nutrition is indeed reflected in the development of this sophisticated water-control technology that since ancient times has impressed explorers and geographers, who refer in their chronicles to astonishing, ingenious plantations (Spriggs 2012). Different theories were proposed regarding who had built them. A simplified view states that taro accompanied the first Austronesian-speaking people, who, sailing eastward across the Pacific, gradually colonised the islands while carrying taro and other crops, such as yam, coconut, and breadfruit, as provisions for their journeys and future settlements (e.g. Horridge 2006).

Pacific islanders are known for having strong maritime traditions, and yet it is their agricultural skill that is evident when it comes to taro cultivation. Throughout the Pacific islands, before contact with Europeans, the use of ploughs or animals was unknown and agricultural work was mainly done with digging sticks (Kirch 1994). Intensified agricultural systems developed on many islands despite the intensive human labour required, probably as an adaptation, for people's economic needs, to restricted mountain environments (Semple 1910, cited in Spriggs 2012).

In the Pacific region, Barrau (1965) was one of the first authors to remark on the dual nature of taro cultivation systems, as summarised in the title of his paper *L'humide et le sec* (the wet and the dry). Taro dry cultivation is typical of southern and northern parts of the tropics, and although 'dry', it is still dependent on seasonal rainfall. The wet cultivation systems include three different forms of cultivation. The most simple is the pit system, typical of atolls, where holes are dug into the ground and plants are placed in a mulched soil close to the freshwater lens located underground. On volcanic islands, raised-bed cultivations, or 'garden islands', are found. These require swamplands, in which raised beds are separated by water channels to ensure adequate drainage. The most complex engineering work for cultivating taro is the pond field irrigation system, still actively used on many islands (e.g. Hawaii, Vanuatu), but largely abandoned in New Caledonia (Sand 2012).

From a social and cultural perspective, the Pacific region offers significant examples of the importance of taro in people's lives. This is particularly true for Pollap, a small coral atoll that is part of the Micronesian Caroline Islands. Here, rainfall creates natural swamps, to which organic material is added to make them suitable for the cultivation of taro. On this island, there is a clear separation of roles; while men are primarily involved in fishing, women work in the taro gardens. In ceremonies initiating new navigators, men cook breadfruit and women cook taro (Flinn 2010). These events seem to be connected to early navigators who brought taro with them during their voyages. The strong relationship that exists between taro and the sea is evident from local stories; one of them relates that after being fed with fish, coconut, and taro by a Pollapese woman, a bird revealed to her the art of navigation (Flinn 2010).

A continuing relationship between women and taro in Pollap is apparent in the present-day cult of the Virgin Mary (Flinn 2010). The eighth day of December is considered to be the most important Catholic festivity. On this day, the Virgin Mary is celebrated and women can display their skills and abilities in growing taro plants with large corms. The church is adorned with taro plants, corms, stalks, and leaves, and flowers mixed with huge taro stalks and leaves decorate the interior, including the altar (Flinn 2010). It is an important moment in women's spiritual lives. The women, offering taro on the feast day, express their gratitude to Mary, who they believe provided them with taro and the skills to cultivate it. Mary is regarded as a nurturer and custodian who provides protection from storms.

On other islands, such as Ifaluk (also in Micronesia), taro is also considered a food of the spirits, with its consumption believed to connect people with the Gods (Flinn 2010). A similar tradition is present on the Hawaiian islands, where the name for taro is *haloa naka*, which literally means 'elder brother'. Taro therefore is important to the family bond (Kahumoku 1990). It is considered holy, and therefore has a meaning beyond that of sustenance.

5. Taro in Africa

Among the root crops that occupy an important role in the modern economy of many African nations are potato (*Solanum tuberosum* L.), cassava (*Manihot esculenta* Crantz), sweet potato (*Ipomea batata* (L.) Lam.), yam (*Dioscorea* spp.), taro (*Colocasia esculenta* (L.) Schott), and yautia (*Xanthosoma sagittifolium* Schott) (FAOstat 2011). Potato, cassava, and yautia were introduced by Europeans from the New World to West Africa to supply food, and as commodity used in exchange for slaves (Carney & Rosomoff 2009). Taro, however, which is native to Asia, appeared in Africa at a time as yet unknown. The primary introduction of this crop from the New World has been ruled out, and over time scholars have put forward several hypotheses to explain its entry into Africa.

Affinities between the Sanskrit term for taro, which is *kutchoo*, and the Greek term *colocasia* led de Candolle (1884) to propose a direct sea-borne introduction of this plant from India or Sri Lanka into the Arabic-speaking world. Of the same opinion was Thiselton-Dyer, who, referring to another Sanskrit name for taro, *kala-katchu*, reported Charles Lyall's arguments for pronounciation: *ch* would be pronounced in western India as *ts* and in northeastern India as *s* (Thiselton-Dyer 1918: 299). Following this semantic path, and using the translation of *kala* as 'black', Portères (1960) put forward the idea that the *black kasia*, or the 'edible black root', originally came from the basin of the Ganges or the Gulf of Bengal. Today, in north-eastern India the name *kala kachu* is still applied to taro (Subhash et al. 2012), highlighting the possibility of a strong connection between this area and the Greek word κολοκασια (latinised as *colocasia*).

Burkill (1935) thought that taro had arrived in Egypt by the time of Christ and that from there it was dispersed across the Mediterranean—although he did not suggest what route it followed and who was responsible for its introduction. He was also convinced that at an early date the plant had been introduced to East Africa through the port of Zanzibar, in modern-day Tanzania, and that from there and also from the Mediterranean the plant was dispersed across Africa.

A sea-borne route was also proposed by Murdock (1959). Following linguistic, ethnographic, and historical evidence, he put forward the idea that during the first century AD, Indonesian people, navigating along the coasts of India, southern Arabia, and East Africa, travelled toward the Azanian coasts, where they introduced the Malaysian crops to Africa. The Malaysian complex included crops of South-East Asian origin that currently grow in Africa, including rice (*Oryza sativa* L.), banana (*Musa* spp.), yam (*Dioscorea* spp.), breadfruit (*Artocarpus altilis* (Parkinson) Fosberg), sugarcane (*Saccharum officinarum* L.), and taro. Murdock suggested that, once in Africa, these crops were transmitted through the 'yam belt', an existing corridor occupied by agricultural people who would have been responsible for the diffusion of the Malaysian crops to the other side of the continent. From equatorial Africa, the plants could spread throughout Africa.

Plucknett (1970, 1976) supported the hypothesis that taro moved with 'Megalithic people' from India to Arabia, Egypt, and the eastern Mediterranean as early as 500 BC (Reinhardt 1911, cited in Plucknett 1970). Then, via the Nile, the plant could have been transported across the continent into West Africa. Noting Pliny's text, he inferred that *colcas, culcas,* or *qolquas* (Arabic: قلقاس) was an important food crop of the Nile Valley and may have been dispersed across Africa by Arabs.

Subscribing to Plucknett's model, and basing his arguments on linguistic evidence, Watson (1983) proposed an Islamic diffusion of taro from the Near East to West Africa

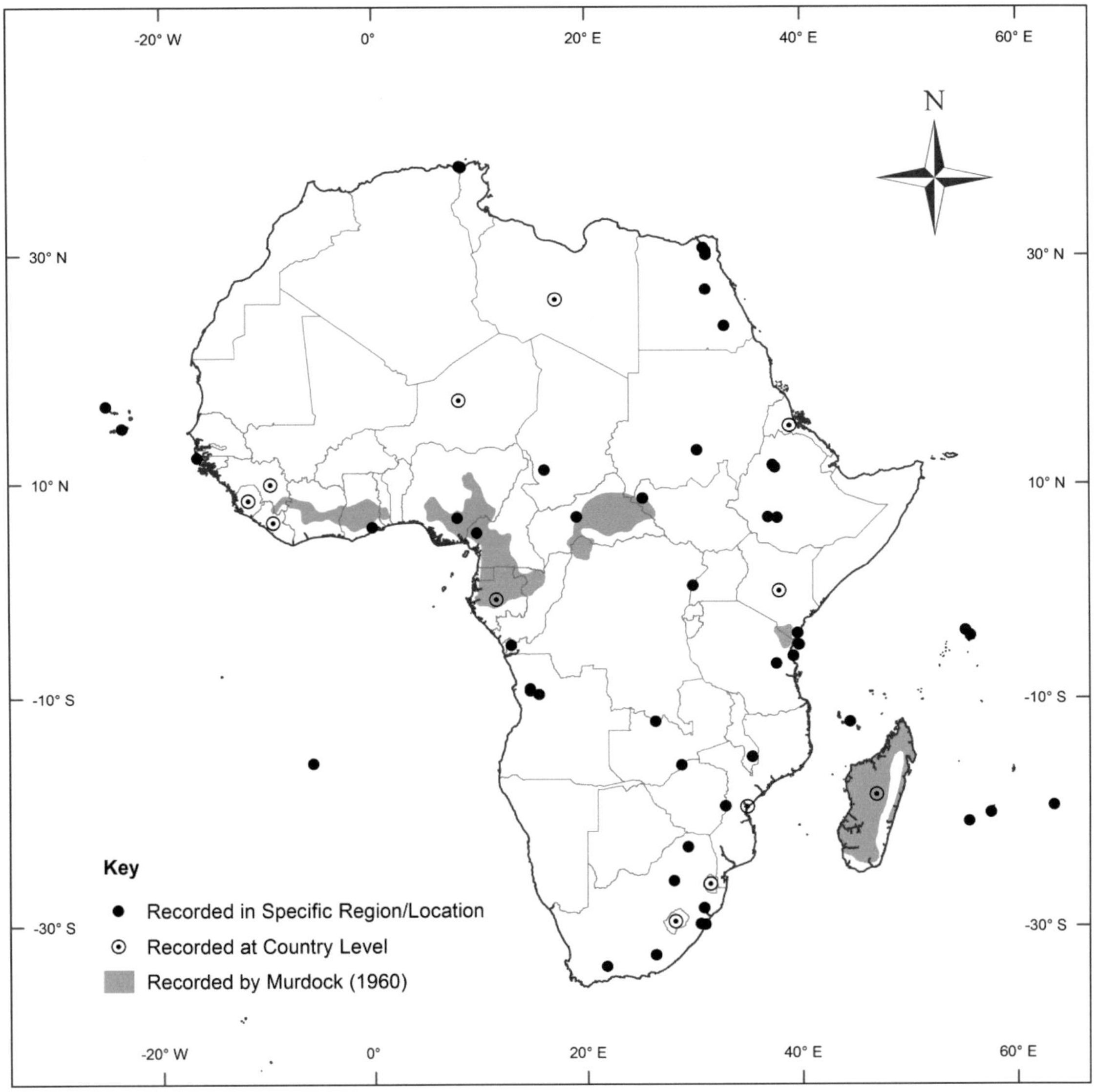

Figure 1. Historic map (1849–2012) of taro distribution, based on the sources listed in Appendix 1. Note: Murdock's data refer only to *Colocasia esculenta* and not to the New World aroid *Xanthosoma sagittifolium*.

via the Nile. In this regard, the recent retrieval of desiccated corms and skin fragments of taro at the Islamic port of Quseir al-Qadim in Egypt provides evidence of a point in time in which taro was part of both Egyptian cuisine and agriculture (Van der Veen & Morales 2011). The identification was based on morphological features of the cut corm fragments, in which typical scars on the corm skin left by the cutting of lateral cormels suggest this was what Purseglove (1972: 61-66) defined as *Colocasia esculenta* var. *antiquorum* (Van der Veen & Morales 2011).

However, written records (Barbier de Meynard & Pavet de Courteille 1864; Portères 1960; Silvatico 1474) seem to indicate that taro was already known in Mesopotamia, the Levant, and Egypt before the tenth century AD, and that by the fourteenth century AD it may well have reached West Africa, where a plant similar to colocasia, known as *kafi*, was recorded by Ibn Batuta in the Kingdom of Mali (Defrémery & Sanguinetti 1858: 399). Linguistic studies in West Africa have dismissed the hypothesis of an Arab introduction in that area (Blench 1997; Williamson 1993), and exactly when and how taro arrived on the shores of the Atlantic remains unknown. The numerous taro cultivars in West Africa and the role this crop occupies in local agriculture and traditional cuisine could indicate a long history in that area. However, strong knowledge of root crop cultivation, such as that of indigenous African yams, may have favoured taro adoption. In this case too, further genetic studies may shed light on the issue.

Typological comparison of musical instruments found in West Africa with those of Asia adds a cultural element

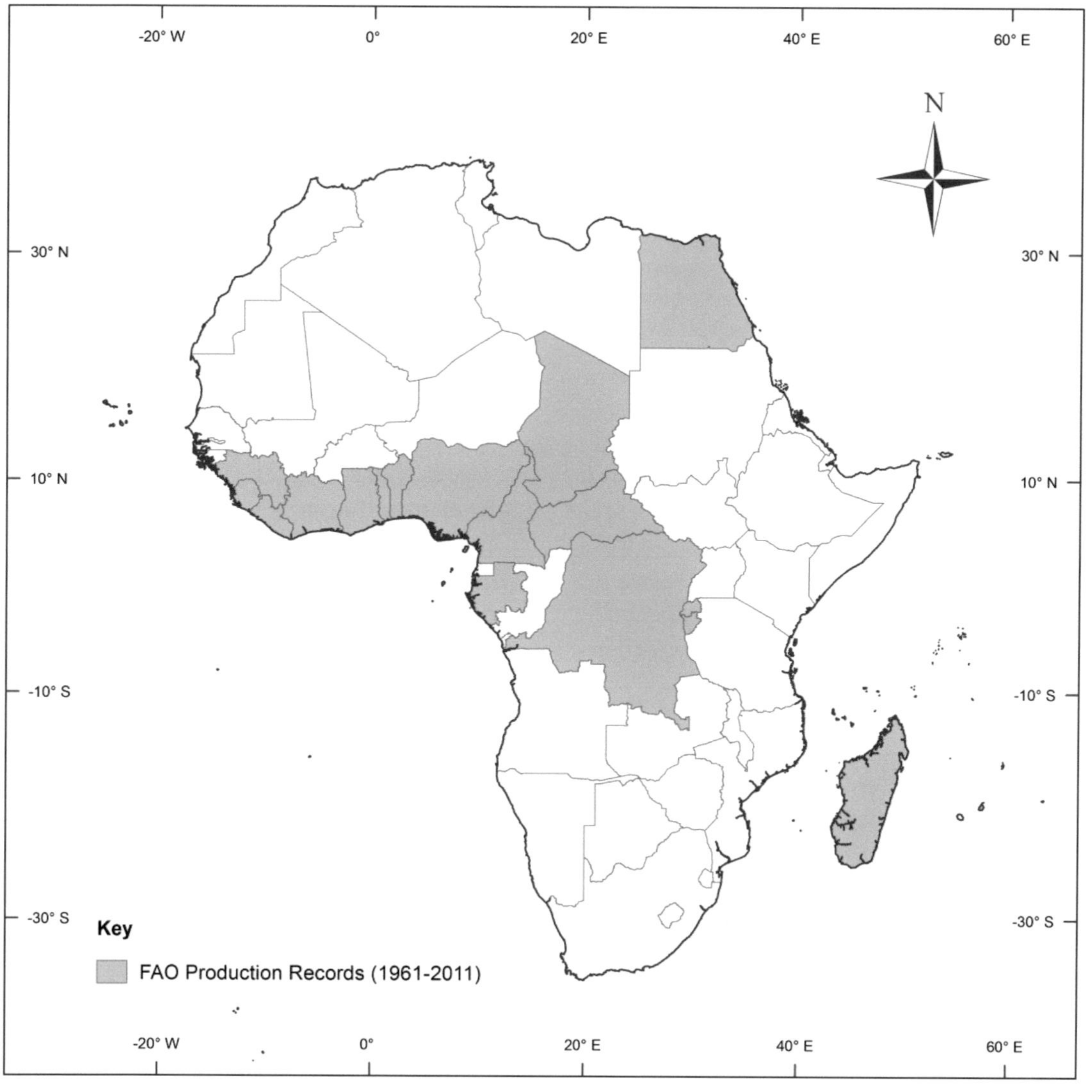

Figure 2. Map of taro production in Africa, based on FAO data recorded from 1961 to 2011 (FAOstat, 2012).

to the discussion. Jones (1971) suggested the hypothesis that Austronesians may have arrived in Madagascar and travelled around the Cape of Good Hope to reach the West African coast. There they would have been responsible for an early introduction of tropical food crops. Although long storage is possible for water yam (*Dioscorea alata* L.) and taro, which lends these crops to long voyages, this model requires people braving difficult sailing conditions to reach West Africa.

In order to visualise the presence of taro in Africa, a distribution map was produced based on records and information found in floristic reports, agronomic reports, and academic papers. Appendix 1 reports all the sources from which data were extracted.

Figure 1 shows the collation of information on taro at the country level as well as in specific locations. A distribution map of taro drawn by Murdock in 1960 has also been overlapped with these data.

A second map was drawn using data collected by the statistic department of the Food and Agriculture Organization of the United Nations (FAOstat 2012) (Fig. 2). This map represents the countries where taro was being cultivated between 1961 and 2011.

These maps show that taro is currently cultivated in hot, humid areas with high rainfall, such as the tropical sub-Saharan region, but that it can also be found along streams in drier areas, such as Algeria, Libya, and Egypt. The comparison between data extrapolated from the FAOstat database and historical records reflects where the production is significant in level and, conversely, where taro is used as an emergency crop or was historically present, and where it thus contributes marginally to the overall crop production of the country.

In West Africa taro is a staple, and the corms and flowers are widely used in the local cuisine to make *achu* or *fufu*, a dish made of mashed corms accompanied by a sauce made with meat, fish, and other vegetables. In this area, research on taro has been carried out in dedicated centres for root crops, such as the Institute of Agricultural Research for Development (IRAD), in Cameroon, and the National Root Crops Research Institute (NRCRI), in Nigeria. Their goal, among others, is to disseminate information on the agronomical and nutritional value of taro and to dispel the idea that taro is a minor crop. In East Africa, taro does not have the same importance as it does in West Africa. In both regions, changes in environment and in agricultural systems generally have led to a decline in taro production, and it is mainly older people who remember old taro cultivars and who bear witness to the introduction of new plants (author's fieldwork observation).

Despite the many recent changes in African agriculture, I expect that analysis of the present distribution of taro over a wide range will provide hints on the original introduction in the distant past.

6. Taro across the Atlantic Ocean

During the African slave diaspora, millions of native African people were removed from their homeland to be transported to the Americas and used as slaves to produce commodities for re-sale in Europe. In this process, people carried ideas, diseases, and cultural identities, in which food represented what Carney & Rosomoff (2009) describe as 'resilience against adversity'.

When the African slave trade commenced, slaveholders, traders, and the slaves themselves had to be fed, and thousands of tons of foodstuff must have been produced to this end. In the sixteenth century, all along the Guinea coasts, slaves were confined in forts prior to being deported across the Atlantic. Provisions for journeys that lasted two or three months had to be collected before departure, and the demand was mostly met using crops grown in Africa. Thus the human trade brought about the development of markets, which supplied food in great quantities for Europeans, African middlemen, and enslaved people alike. Human trafficking radically changed food production strategies. In America, initiatives were undertaken by the slaves themselves to plant crops for their own consumption, and provisions that occasionally remained from slave voyages may have been a source of planting materials.

The production of sugar, initially in the Mediterranean and Africa and subsequently in the Americas, also opened up routes for other crops, such as coffee, tobacco, cotton, and rice. African yam and plantains had a bigger impact on the New World tropical food system (Carney & Rosomoff 2009), but taro also crossed the Atlantic and became widespread in the New World. The image of ships crowded with slaves and crops stored on board is far from romantic; however, this was how many plants crossed the Atlantic, giving sustenance—and perhaps hope—to the migrants.

Many of the plants carried across the Atlantic were unknown to Europeans and consequently were given generic and new names. Sometimes crops were named after their place of last origin, as in the names guinea squash, guinea melon, da costa, or da terra, to indicate that they were from the coast or from inland. Taro became known by the common name inhame da costa, together with African yam, *Dioscorea* spp. (Carney & Rosomoff 2009).

There seems to be no specific record for the introduction of taro in the Americas, part of the problem being that yautia and taro are similar in appearance, and often have collective names (Burkill 1938). Richard Ligon (1657) noticed that in Barbados, African people cultivated roots that grew from 'seeds' (i.e. planting stocks) that the slaves had brought with them on the voyage across the ocean. His description given for this crop—plants with big leaves, which are eaten along with the roots and which are suitable for long voyages—suggests that he was referring to taro (Carney & Rosomoff 2009). Linnaeus (1753) described *Arum esculentum* from specimens of taro collected in America, and it therefore seems reasonable to think that the New World received taro first from West Africa and later, in recent times, from the Pacific. Cook (1910) noted that Chinese residents in California imported taro from Canton, China, and Hawaii and that the vast wet uncultivated areas in the Florida Gulf and southern Atlantic states could be exploited to grow the numerous varieties of taro and other edible aroids.

7. Taro across the world

In our hyper-connected world, plants are now traded and moved on a global scale. When taro first crossed oceans long ago, translocations were entirely dependent on people accompanying it from one continent to another.

Early domestication and cultivation of taro in Asia presumably facilitated dispersal of the plant across the Pacific with agriculturist mariners. Even though it is not possible yet to establish a date for its arrival in Africa, taro could certainly have been transported across the Indian Ocean. The relationship with the sea was instrumental, and its use as food for long transatlantic journeys during the African slave trade made it part of 'Africa's botanical legacy in the Atlantic World' (Carney & Rosomoff 2009). Once taro was carried across the Pacific to the western seaboard of the Americas, it had finally completed a global journey that lasted thousands of years.

Acknowledgments

This author would like to thank Dr Nicole Boivin, Principal Investigator of the SEALINKS Project (University of Oxford, UK); Dr Robin G. Allaby (University of Warwick, UK); the European Research Council; the Meyerstein Fund (School of Archaeology, University of Oxford); and Wolfson College (University of Oxford) for supporting this research and its presentation at the 7th International Workshop for African Archaeobotany. The author is also grateful to Dr Peter J. Matthews (National Museum of Ethnology, Osaka, Japan) for his comments and for editing the English in early drafts.

References

Ahmed I., Biggs P.J., Matthews P.J., Collins L.J., Hendy M.D. & Lockhart P.J., 2012. Mutational dynamics of aroid chloroplast genomes. *Genome Biology and Evolution* 4 (12), pp. 1316-1323.

Andrews F.W., 1956. *The Flowering Plants of the Sudan.* Vol. 3. Compositae–Gramineae. Published on behalf of the Sudan Government by T. Buncle and Co., Arbroath.

Anguillara L.M., 1561. *Semplici, li quali in più parerei a diversi nobili huomini scritti appaiono.* A cura di G. Marinello, Venezia. pp. 226-228.

Antonsson-Ogle B., 1984. Wild plant resources, another form of food security. *CERES* 17 (4), pp. 38-40.

Bandyopadhyay R., Sharma K., Onyeka T.J., Aregbesola A. & Kumar P.L., 2011. First report of taro (*Colocasia esculenta*) leaf blight caused by *Phytophthora colocasiae* in Nigeria. *Plant Disease* 95, p. 618.

Barbier de Meynard C. & Pavet de Courteille A. (eds. and transl.), 1864. *Al-Masudi Les prairies d'or.* Vol. 3. [Original work written c. 10th century AD.] Imprimerie imperiale, Paris.

Barrau J., 1965. L'Humide et le sec: An essay on ethnobiological adaptation to contrastive environments in the Indo-Pacific area. *Journal of the Polynesian Society* 74 (3), pp. 329-346.

Battandier J.A. & Trabut L., 1885. *Flore d'Alger et catalogue des plantes d'Algérie.* Adolphe Jourdan, Alger.

Bellwood P., 2006. Austronesian prehistory in southeast Asia: Homeland, expansion and transformation. In: Bellwood P., Fox J.J. & Tryon D. (eds.), *The Austronesians: Historical and Comparative Perspectives.* Australian National University Press, Canberra, pp. 103-118.

Berhaut J., 1988. *Flore illustrée du Sénégal.* Vol. 9. Monocotylédones et ptéridophytes. Gouvernement du Sénégal, Ministère de la protection de la nature, Direction des eaux et forêts, Dakar.

Blench R.M., 1997. A history of agriculture in northeastern Nigeria. In: Barreteau D., Dognin R. & von Graffenried C. (eds.), *L'Homme et le milieu végétal dans le bassin du Lac Tchad.* ORSTOM, Paris, pp. 69-112.

Blench R.M., 2012. Vernacular names for taro in the Indo-Pacific Region: Implications for centres of diversification and spread. In: Spriggs M., Addison D. & Matthews P.J. (eds.), *Irrigated Taro (*Colocasia esculenta*) in the Indo-Pacific: Biological, Social and Historical Perspectives.* Senri Ethnological Studies 78, Osaka, pp. 21-43.

Bogner J., 1975. *Flore de Madagascar et des Comores.* 31e famille. Aracées. Museum national d'histoire naturelle, Paris.

Bosser J., 1976. *Flore des Mascareignes : La Réunion, Maurice, Rodrigues.* MSIRI, Mauritius; ORSTOM, Paris; Royal Botanic Gardens, Kew.

Broun A.F. & Massey R.E., 1929. *Flora of the Sudan.* Willington House, London.

Burkill I.H., 1935. *A Dictionary of the Economic Products of the Malay Peninsula.* Vol. 1. Published on behalf of the governments of the Straits Settlements and Federated Malay States by the Crown Agents for the Colonies, London.

Burkill I.H., 1938. The contact of the Portuguese with African food plants which gave words such as 'yam' to European languages. *Proceedings of the Linnean Society of London* 150 (2), pp. 84-95.

Burkill H.M. 1988. *The useful plants of west tropical Africa.* 2nd ed. Royal Botanic Gardens, Kew.

Cabrera L.I., Salazar G.A., Chase M.W., Mayo S.J., Bogner J. & Dávila P., 2008. Phylogenetic relationships of aroids and duckweeds (Araceae) inferred from coding and noncoding plastid DNA. *American Journal of Botany* 95 (9), pp. 1153-1165.

Candolle, A. de, 1884. *Origin of Cultivated Plants.* Kegan Paul, Trench & Co., London.

Cao L.M. & Long C.L., 2003. *Colocasia bicolor* (Araceae), a new species from Yunnan, China. *Annales Botanici Fennici* 40 (4), pp. 283-286.

Caillon S., Quero-Garcîa J., Lescure J.P. & Lebot V., 2006. Nature of taro (*Colocasia esculenta* (L.) Schott) genetic diversity prevalent in a Pacific Ocean island, Vanua Lava, Vanuatu. *Genetic Resources and Crop Evolution* 53 (6), pp. 1273-1289.

Carney J.A. & Rosomoff R.N., 2009. *In the Shadow of Slavery: Africa's Botanical Legacy in the Atlantic World*. University of California Press, Berkeley and Los Angeles.

Chevalier A., 1913. *Études sur la flore de l'Afrique centrale française (bassins de l'Oubanguied du Chari)*. Vol. 1. Énumération des plantes récoltées. A. Challamel, Paris.

Chevalier A. & Chevalier A., 1935. *Les îles du Cap Vert : flore de l'archipel*. Muséum national d'histoire naturelle, Laboratoire d'agronomie coloniale, Paris.

Coates D.J., Yen D.E. & Gaffey P.M., 1988. Chromosome variation in taro, *Colocasia esculenta*: Implications for origin in the Pacific. *Cytologia* 53 (3), pp. 551-560.

Cook O.F., 1910. Agricultural history and utility of the cultivated aroids. *Bulletin Bureau of Plant Industry* (United States Department of Agriculture) 164, pp. 31-37.

Cusimano N., Bogner J., Mayo S.J., Boyce P.C., Wong S.Y., Hesse M. & French J.C., 2011. Relationships within the Araceae: Comparison of morphological patterns with molecular phylogenies. *American Journal of Botany* 98 (4), pp. 654-668.

Dalziel J.M., 1937. *Useful Plants of West Tropical Africa*. Crown Agent for Overseas Governments and Administrations, London.

Defrémery C. & Sanguinetti B.R., 1858. Ibn Batuta *Voyages d'Ibn Batoutah*. Vol. 4. [Arabic text with a translation in French.] Imprimerie impériale, Paris.

Denham T., 2010. From domestication histories to regional prehistory: Using plants to re-evaluate early and mid-Holocene interaction between New Guinea and Southeast Asia. *Food and History* 8 (1), pp. 3-22.

Doku E.V., 1967. Root crops in Ghana. *Journal of Tropical Geography* 11 (19), p. 9.

Durand T. & Schinz H., 1895. *Conspectus floræ Africæ : ou énumération des plantes d'Afrique*. Vol. 5. Jardin botanique de l'état, Brussels.

Engels J.M., Hawkes J.G. & Worede M. (eds.). 1991. *Plant Genetic Resources of Ethiopia*. Cambridge University Press, Cambridge.

FAOstat, 2011. Available at: http://faostat3.fao.org/home/index.html#DOWNLOAD (accessed 5 March 2013).

FAOstat, 2012. Available at: http://faostat3.fao.org/faostat-gateway/go/to/download/Q/QC/E (accessed 17 September 2013).

Flinn J., 2010. *Mary, the Devil, and Taro: Catholicism and Women's Work in a Micronesian Society*. University of Hawaii Press, Honolulu.

Francis C.A., Davidson D.J., Massaquoi R.C., Mulbah C.K., Massaquoi W.K. & Beebe J., 1995. Designing an integrated cropping systems research program: Central Agricultural Research Institute (CARI), Liberia. *Journal of Sustainable Agriculture* 5 (3), pp. 147-168.

Friis I. & Vollesen K., 1998. *Flora of the Sudan–Uganda Border Area East of the Nile*. Danske Videnskabernes Selskab, Copenhagen.

Fullagar R., Field J., Denham T. & Lentfer C., 2006. Early and mid Holocene tool-use and processing of taro (*Colocasia esculenta*), yam (*Dioscorea* sp.) and other plants at Kuk Swamp in the highlands of Papua New Guinea. *Journal of Archaeological Science* 33, pp. 595-614.

Fujimoto T., 2009. Taro (*Colocasia esculenta* (L.) Schott) cultivation in vertical wet–dry environments: Farmers' techniques and cultivar diversity in southwestern Ethiopia. *Economic botany* 63 (2), pp. 152-166.

Govaerts R. & Frodin D.G., 2002. *World Checklist & Bibliography of Araceae (and Acoraceae)*. Royal Botanic Gardens, Kew.

Guarino L., 2010. Taro leaf blight in Cameroon? Agricultural Biodiversity Weblog. Available at: http://agro.biodiver.se/2010/07/taro-leaf-blight-in-cameroon/ (accessed 6 November 2012).

Hammer K., Lehmann C.O. & L'errino P., 1988. A check-list of the Libyan cultivated plants including an inventory of the germplasm collected in the years 1981, 1982 and 1983. *Die Kulturpflanze* 36 (3), pp. 475-527.

Henderson L., 2007. Invasive, naturalized and casual alien plants in southern Africa: A summary based on the Southern African Plant Invaders Atlas (SAPIA). *Bothalia* 37 (2), pp. 215-248.

Hill A.F., 1939. The nomenclature of the taro and its varieties. *Botanical Museum Leaflets* 7 [Harvard University], pp. 113-118.

Hilu K.W., 1993. Polyploidy and the evolution of domesticated plants. *American Journal of Botany* 80 (12), pp. 1494-1499.

Holland J.H., 1922. *The useful plants of Nigeria*. Kew Bulletin Additional Series 9. Her Majesty's Stationery Office, London.

Horridge A., 2006. The Austronesian conquest of the sea—upwind. In: Bellwood P., Fox J.J. & Tryon D. (eds.), *Austronesians, Historical and Comparative Perspectives*. Australian National University Press, Canberra, pp. 143-160.

Hotta M., 1970. *A System of the Family Araceae in Japan and Adjacent Areas*. Part 1. Memoirs of the Faculty of Science, Kyoto University Series of Biology 4, Kyoto, pp. 72-96.

Hotta M., 2002. The origins and spread of tuber crops (*Imo*). In: Yoshida S. & Matthews P.J. (eds.), *Vegeculture in Eastern Asia and Oceania*. Japan Centre for Area Studies, Osaka, pp. 17-30.

Huang S., 2012. Yu Jing (The Book of Taro) In: Spriggs M., Addison D. & Matthews P.J. (eds.), *Irrigated Taro (*Colocasia esculenta*) in the Indo-Pacific: Biological, Social and Historical Perspectives.* Senri Ethnological Studies 78, Osaka, pp. 45-51.

International Plant Names Index. 2014. http://www.ipni.org/

Irvine F.R., 1930. *Plants of the Gold Coast.* Oxford University Press, Humphrey Milford, London.

Irwin S.V., Kaufusi P., Banks K., De la Pena R. & Cho J.J., 1998. Molecular characterization of taro (*Colocasia esculenta*) using RAPD markers. *Euphytica* 99 (3), pp. 183-189.

Isshiki S., Otsuka K., Tashiro Y. & Miyazaki S., 1999. A probable origin of triploids in taro (*Colocasia esculenta* (L.) Schott). *Journal Japanese Society for Horticultural Science* 68, pp. 774-779.

Jones A.M., 1971. *Africa and Indonesia: The Evidence of the Xylophone and Other Musical Factors.* Brill, Leiden.

Kahumoku G., Jr., 1990. A Hawaiian perspective on taro growing. In: Hollyer J.R. & Sato D.M. (eds.), *Proceedings of Taking Taro into the 1990s: A Taro Conference.* University of Hawaii, Honolulu, pp. 88-89.

Kirch P.V., 1994. *The Wet and the Dry: Irrigation and Agricultural Intensification in Polynesia.* University of Chicago Press, Chicago and London.

Koechlin J., Guillaumet J.L. & Morat P., 1974. Flore et végétation de Madagascar. In: Tüxen R. (ed.), *Flora et vegetatio mundi.* J. Cramer, Vaduz.

Kreike C.M., Van Eck H.J. & Lebot V., 2004. Genetic diversity of taro, *Colocasia esculenta* (L.) Schott, in Southeast Asia and the Pacific. *Theoretical and Applied Genetics* 109 (4), pp. 761-768.

Kuruvilla K.M. & Singh A., 1981. Karyoptic and electrophoretic studies on taro and its origin. *Eyphytica* 30 (2), pp. 405-413.

Lebot V. & Aradhya K.M., 1991. Isozyme variation in taro (*Colocasia esculenta* (L.) Schott) from Asia and Oceania. *Euphytica* 56, pp. 55-66.

Lebot V., Prana M.S., Kreike,N., Van Heck H., Pardales J., Okpul T. & Yap T. C., 2004. Characterisation of taro (*Colocasia esculenta* (L.) Schott) genetic resources in southeast Asia and Oceania. *Genetic Resources and Crop Evolution* 51 (4), pp. 381-392.

Lebot V., 2009. *Tropical root and tuber crops: Cassava, sweet potato, yams, aroids.* Crop Production Science in Horticulture 17. CABI, Wallingford.

Léon J., 1977. Origin, evolution and early dispersal of root and tuber crops. In: Cock L., MacIntyre R. & Graham M. (eds.), *Proceedings of the Fourth Symposium of the International Society for tropical Root Crops.* International Development Research Centre, Ottawa, pp. 20-36.

Ligon R., 1657. *A true & exact history of the island of Barbados illustrated with a mapp of the island, as also the principall trees and plants there, set forth in their due proportions and shapes, drawne out by their severall and respective scales together with the ingenio that makes the sugar, with the plots of the severall houses, roomes, and other places that are used in the whole processe of sugar-making by Richard Ligon, Gent.* Humphrey Moseley, London.

Linnaeus C., 1753. *Species Plantarum.* 2 vols. Laurentius Salvius, Stockholm.

Loy T.H., Spriggs M. & Wickler S., 1992. Direct evidence for human use of plants 28,000 years ago: Starch residues on stone artefacts from the northern Solomon Islands. *Antiquity* 66, pp. 898-912.

Mace E.S. & Godwin I.D., 2002. Development and characterization of polymorphic microsatellite markers in taro (*Colocasia esculenta*). *Genome* 45 (5), pp. 823-832.

Mace E.S., Mathur P.N., Godwin I.D., Hunter D., Taylor M.B., Singh D., DeLacy I.H. & Jackson G.V.H., 2006. Development of a regional core collection (Oceania) for taro, *Colocasia esculenta* (L.) Schott, based on molecular and phenotypic characterization. In: Rao R.V., Matthews P.J. & Eyzaguirre P.B. (eds.), *The Global Diversity of Taro: Ethnobotany and Conservation.* Bioversity International, Rome, pp. 185-201.

Mace E.S., Mathur P.N., Izquierdo L., Hunter D., Taylor M.B., Singh D. & Godwin I. D., 2006. Rationalization of taro germplasm collections in the Pacific Island region using simple sequence repeat (SSR) markers. *Plant Genetic Resources: Characterization and Utilization* 4 (3), pp. 210-220.

Marloth R., 1915. *The Flora of South Africa: With Synoptical Tables of the Genera of the Higher Plants.* Vol. 4. Darter bros. and Co., Cape Town, pp. 69-72.

Matsuda M. & Nawata E., 2002. Geographical distribution of ribosomal DNA variation in taro, *Colocasia esculenta* (L.) Schott, in eastern Asia. *Euphytica* 128 (2), pp. 165-172.

Matthews P.J., 1990. *The Origins, Dispersal and Domestication of Taro.* Unpublished Ph.D. Thesis, Australian National University, Canberra.

Matthews P.J., 1991. A possible tropical wildtype taro: *Colocasia esculenta* var. *aquatilis. Indo-Pacific Prehistory Association Bulletin* 11, pp. 69-81.

Matthews P.J., Takei E. & Kawahara T., 1992. *Colocasia esculenta* var. *aquatilis* on Okinawa Island, southern Japan: The distribution and possible origins of a wild diploid taro. *Man and Culture in Oceania* 8, pp. 19-34.

Matthews P.J. & Terauchi R., 1994. The genetics of agriculture: DNA variation in taro and yam. In: Hathe J.G. (ed.), *Tropical Archaeobotany: Applications and New Developments.* Routledge, London, pp. 251-270.

Matthews P.J., 1995. Aroids and the Austronesians. *Tropics* 4 (2), pp. 105-126.

Matthews P.J., 1998. Wildtype taro, and the recent history of cultivated taro in Cyprus and Hawaii. In: Agricultural and Forestry Research Center (ed.), *Application of Biological Resources for the Innovation of Agricultural and Environmental Education in Asian-Pacific Countries.* [Proceedings of 1998 Tsukuba Asian Seminar on Agricultural Extension.] Agricultural and Forestry Research Center, University of Tsukuba, Tsukuba, pp. 99-108.

Matthews P.J., 2005. Notes on the provenance and providence of wildtype taros (*Colocasia esculenta*) in Myanmar. *Bulletin of the National Museum of Ethnology* [Japan] 29 (4), pp. 587-615.

Mayo S.J., 1985. Araceae. In: Polhill R.M. (ed.), *Flora of Tropical East Africa.* Published on behalf of the East African Government by Balkema, Rotterdam.

Mengistu M.G., Everson C.S. & Clulow A.D., 2012. The impact of taro (*Colocasia esculenta*) cultivation on the total evaporation of a *Cyperus latifolius* marsh. *Hydrological Process* 28 (3), pp. 620-627.

Murdock G.P., 1959. *Africa: Its People and Culture History.* McGraw-Hill, New York.

Murdock, G.P., 1960. Staple subsistence crops of Africa. *Geographical Review* 50 (4), pp. 523-540.

Nauheimer L., Boyce P.C. & Renner S.S., 2012. Giant taro and its relatives: A phylogeny of the large genus *Alocasia* (Araceae) sheds light on Miocene floristic exchange in the Malesian region. *Molecular Phylogenetics Evolution* 63 (1), pp. 43-51.

Nguyen V.X., Yoshino H. & Tahara M., 1998. *Phylogenetic Analyses of Taro (*Colocasia esculenta *(L.) Schott) and Related Species based on Esterase Isozymes.* Scientific Reports, Faculty of Agriculture, Okayama University 87, pp. 133-140.

Ntepe-Nyame C., 1988. *Araceae : flore du Cameroun.* Ministère de l'enseignement supérieur, de l'informatique et de la récherche scientifique, Yaoundé.

Okonkwo C.A.C., 1993. Taro *Colocasia* spp. In: Kalloo G. & Bergh G. (eds.), *Genetic Improvement of Vegetable Crops.* Pergamon, Oxford, pp. 709-715.

Oliveira J.S. & De Carvalho M.F., 1975. Nutritional value of some edible leaves used in Mozambique. *Economic Botany* 29 (3), pp. 255-263.

Omane E., Oduro K.A., Cornelius E.W., Opoku I.Y., Akrofi A.Y., Sharma K., Kumar P.L. & Bandyopadhyay R., 2012. First report of leaf blight of taro (*Colocasia esculenta*) caused by *Phytophthora colocasiae* in Ghana. *Plant Disease* 96 (2), p. 292.

Parvin S., Kabir G., Ud-Deen M.M. & Sarker J.K., 2008. Karyotype analysis of seven varieties of taro *Colocasia esculenta* (L.) Schott. from Bangladesh. *Journal of Bio-Science* 16, pp. 15-18.

Plucknett D.L., 1970. Tropical root and tuber crops tomorrow: *Colocasia, Xanthosoma, Alocasia, Cyrtosperma,* and *Amorphophallus.* In: Plucknett D.L. (ed.), *Proceedings of the Second International Symposium on Tropical Root and Tuber Crops.* College of Tropical Agriculture, University of Hawaii, Honolulu, pp. 127-135.

Plucknett D.L., 1976. Edible aroids. In: Simmonds N. (ed.), *Evolution of Crop Plants.* Longman Group, London, pp. 10-12.

Plucknett D.L., 1983. Taxonomy of the genus *Colocasia.* In: Wang J.-K. (ed.), *Taro, a Review of* Colocasia esculenta *and Its Potentials.* University of Hawaii Press, Honolulu, pp. 14-19.

Pobéguin C.H.O., 1906. *Essai sur la flore de la Guinée Française : produits forestiers, agricoles et industriels.* A. Challamel, Paris.

Portères R., 1960. La sombre Aroidée cultivée : *Colocasia antiquorum* (L. Schott) ou du taro de Polynésie, essai d'etymologie sémantique. *Journal d'agriculture tropicale et de botanique appliqueé* 7, pp. 169-190.

Porterfield W.M., Jr., 1944. Taro, ancient food plant of tropical regions. *Journal of New York Botanical Garden* 45, pp. 11-16.

Purseglove J.W., 1972. Araceae. In: *Tropical Crops.* Vol 1. Monocotyledons. Longman Group, London. pp. 58-74.

Quézel P., Santa S. & Schotter O., 1962. *Nouvelle flore de l'Algérie et des regions désertiques méridionales.* Vol. 1. Éditions du Centre national de la récherche scientifique, Paris.

Ramsar, 2005. *The Annotated Ramsar List of Wetlands of International Importance.* http://www.ramsar.org/cda/en/ramsar-pubs-notes-anno-chad/main/ramsar/1-30-168%5E16481_4000_0__ (accessed 20 May 2013).

Reinhardt L., 1911. *Kulturgeschichte der Nutzpflanzen.* Ernst Reinhardt, Munich.

Rendle A.B., 1899. *Catalogue of the African plants.* Vol. 2. Part 1. Printed by order of the Trustees of the British Museum (Natural History).

Renner S.S. & Zhang L.B., 2004. Biogeography of the Pistia clade (Araceae): Based on chloroplast and mitochondrial DNA sequences and Bayesian divergence time inference. *Systematic Biology* 53 (3), pp. 422-432.

Republic of South Africa, Department of Agriculture, Forestry & Fisheries. 2011. *Production Guidelines for Amadumbe.* Department of Agriculture, Forestry & Fisheries, Pretoria. http://www.daff.gov.za/docs/Brochures/Amadumbe_1.pdf

Sacy S. de, 1810. *Relation de l'Égypte par Abd-Allatif, médecin arabe de Bagdad.* Imprimerie impériale, Paris.

Safo Kantanka O., 2004. *Colocasia esculenta* (L.) Schott. In: Grubben G.J.H. & Denton O.A. (eds.), *Plant Resources of Tropical Africa.* Vol. 2. Vegetables. Backhuys, Leiden, pp. 206-211.

Sand C., 2012. 'Certainly the most technically complex pond-field irrigation within Melanesia': Wet taro field systems of New Caledonia. In: Spriggs M., Addison D. & Matthews P.J. (eds.), *Irrigated Taro (*Colocasia esculenta*) in the Indo-Pacific: Biological, Social and Historical Perspectives.* Senri Ethnological Studies 78, Osaka, pp. 167-188.

Schott H.W., 1832. Araceae. In: Schott H.W. & Endlicher S. (eds.), *Meletemata Botanica*. Vol. 1. C. Gerold, Vienna, p. 18.

Schott H.W., 1856. *Synopsis Aroidearum Complectens: Enumerationem Systematicum Generum et Specierum Hujus Ordinis.* Vienna.

Semple E.C., 1910. *Influences of Geographic Environment on the Basis of Ratzel's System of Anthropo-geography*. Henry-Holt, New York.

Silvatico M., 1474. *Liber pandectarum medicinae (opus emendatum per ... dominum Matheum Moretum)*. I. Vurster, Bologna.

Singh D., Mace E.S., Godwin I.D., Mathur P.N., Okpul T., Taylor M. & Jackson G., 2008. Assessment and rationalization of genetic diversity of Papua New Guinea taro (*Colocasia esculenta*) using SSR DNA fingerprinting. *Genetic Resources and Crop Evolution* 55 (6), pp. 811-822.

Singh D., Jackson G., Hunter D., Fullerton R., Lebot V., Taylor M. & Tyson J., 2012. Taro Leaf Blight—A Threat to Food Security. *Agriculture* 2 (3), pp. 182-203.

Spriggs M., 2012. From Mendana to Riesenfeld: Early accounts of and speculation on taro irrigation in the Asia-Pacific area. In: Spriggs M., Addison D. & Matthews P.J. (eds.), *Irrigated Taro* (Colocasia esculenta) *in the Indo-Pacific: Biological, Social and Historical Perspectives*. Senri Ethnological Studies 78, Osaka, Japan, pp. 1-19.

Subhash M., Manab D., Pratim S.M., Minakshi B., Md G.I. & Giasuddin A., 2012. High content of aflatoxin in the indigenous foods as possible risk factor for liver cancer: An experience from Assam, India. *Journal of Environmental Research and Development* 7(2a), pp. 838-843.

Täckholm V. & Drar M., 1941. *The Flora of Egypt*. Vol. 2. Cairo University Press, Cairo.

Taylor M., Hunter D., Rao V.R., Jackson G.V.H., Sivan P. & Guarino L., 2010. Taro collecting and conservation in the Pacific Region. In: Rao V.R., Matthews P.J., Eyzaguirre P.B. & Hunter D. (eds.), 2010. *The Global Diversity of Taro: Ethnobotany and Conservation*. Biodiversity International, Rome, pp. 150-167.

The Plant List. 2013. Version 1.1. http://www.theplantlist.org/

Thiselton-Dyer W.T., 1918. On some ancient plants. *Journal of Philology* 34, pp. 299-304.

Van der Veen M. & Hamilton D., 2011. Chronology. In: Van der Veen M. (ed.), *Consumption, Trade and Innovation: Exploring the Botanical Remains from the Roman and Islamic Ports at Quseir al-Qadim, Egypt*. Journal of African Archaeology Monograph Series 6. Africa Magna Verlag, Frankfurt, pp. 18-31 plus Appendix 1.

Van der Veen M. & Morales J., 2011. Summer crops—from trade to innovation. In: Van der Veen M. (ed.), *Consumption, Trade and Innovation: Exploring the Botanical Remains from the Roman and Islamic Ports at Quseir al-Qadim, Egypt*. Journal of African Archaeology Monograph Series 6. Africa Magna Verlag, Frankfurt, pp. 75-119.

van Rensburg J.W.S., Shanahan P. & Greyling M.M., 2011. Molecular characterization of South African taro (*Colocasia esculenta*) germplasm using SSR markers. In: *Crops for the Future—Beyond Food Security*. [Proceedings of the Second International Symposium on Underutilized Plant Species.] ISHS Acta Horticulturae 979, pp. 459-466.

Vogel T., Webb P.B., Hooker J.D. & Bentham G., 1849. *Niger flora: or, An enumeration of the Plants of Western Tropical Africa*. Vol. 1. H. Baillière, London.

Walsh M., 2009. The use of wild and cultivated plants as famine foods on Pemba Island, Zanzibar. *Études Océan Indien* 42–43, pp. 217-241.

Watson A.M., 1983. *Agricultural Innovation in the Early Islamic World*. Cambridge University Press, Cambridge.

Williams R.O., 1949. *The Useful Ornamental Plants of Zanzibar and Pemba*. St. Ann's Press, Zanzibar.

Williamson K., 1970. Some food plant names in the Niger Delta. *International Journal of American Linguistics* 36, pp. 156-167.

Williamson K., 1993. Linguistic evidence for the use of some tree and tuber food plants in southern Nigeria. In: Shaw T., Sinclair P., Andah B. & Okpoko A. (eds.), *The Archaeology of Africa: Food, Metals and Towns*. Routledge, London, pp. 104-116.

Wilson J.E., 1979. Promotion of flowering and production of seed in cocoyam (*Xanthosoma* and *Colocasia*). In: *Proceedings of the V International Symposium on Tropical Root and Tuber Crops*. Provisional Report of the International Foundation for Science 5, pp. 269-279.

Yen D.E. & Wheeler J.M., 1968. Introduction of Taro into the Pacific: The indications of the chromosome numbers. *Ethnology* 7 (3), pp. 259-267.

Yin J.T., Li H. & Xu Z.F., 2004. *Colocasia menglaensis* (Araceae), a new species from southern Yunnan, China. *Annales Botanici Fennici* 41 (3), pp. 223-226.

Yin J., 2006. *Colocasia tibetensis* (Araceae, Colocasieae), a new species from southeast Tibet, China. *Annales Botanici Fennici* 43 (1), p. 53.

Yoshino H., 2002. Morphological and genetic variation in cultivated and wild taro. In: Shuji Y. & Matthews P.J. (eds.), *Vegeculture in Eastern Asia and Oceania*. [Proceedings of the Seventh International Area Studies Conference.] Japan Centre for Area Studies Symposium Series 16, National Museum of Ethnology, Osaka, pp. 95-116.

Young R.A., 1917. *The dasheen; Its uses and culture*. Yearbook of the Department of Agriculture [USA], 1916.

Appendix 1

Sources for historical and contemporary records of taro distribution in Africa.

Country	Region/location	Type of reference	Publication
Algeria	Cap Rosa	Floristic	Battandier & Trabut 1885
		Floristic	Quézel et al. 1962
Angola		Floristic	Durand & Schinz 1895
	Golungo Alto, Cazengo, Pungo Andongo	Floristic	Rendle 1899
Cameroon		Floristic	Ntepe-Nyame 1988
Cape Verde	S. Antão, S. Thiago, Orgãos		Chevalier & Chevalier 1935
		Floristic	Täckholm & Drar 1941
Central Africa	Béreki		Chevalier 1913
Chad			Ramsar 2005
Comoros			Bosser 1976
Congo			Täckholm & Drar 1941
Côte d'Ivoire			Burkill 1988
Egypt			Täckholm & Drar 1941
Ethiopia		Floristic	Durand & Schinz 1895
	Malo (Gaytsa, Shaasha, Borodda, Ziita, Falaha, Haste)	Agricultural report	Fujimoto 2009
Eritrea			Engels et al. 1991
Gabon			Wilson 1979
Ghana	Akropong, Akwapim	Agricultural report	Doku 1967
		Floristic	Irvine 1930
Gold Coast		Floristic	Täckholm & Drar 1941
Guinea		Floristic	Pobéguin 1906
Kenya		Floristic	Täckholm & Drar 1941
	East Africa	Herbarium	Mayo 1985
La Réunion		Floristic	Bosser 1976
			Täckholm & Drar 1941
Lesotho		Ethnographic	Henderson 2007
Liberia		Agricultural report	Francis et al. 1995
Libya		Floristic	Hammer et al. 1988
Madagascar		Floristic	Täckholm & Drar 1941
		Floristic	Koechlin et al. 1974
		Floristic	Bogner 1975
		Floristic	Bosser 1976
Mauritius		Floristic	Durand & Schinz 1895
		Floristic	Täckholm & Drar 1941
		Floristic	Bosser 1976
Mozambique		Agricultural report	Oliveira & De Carvalho 1975
Niger		Floristic	Vogel et al. 1849
Bas-Niger		Floristic	Durand & Schinz 1895
Nigeria		Agricultural report	Holland 1922
			Täckholm & Drar 1941
Rodrigues			Bosser 1976

Country	Region/location	Type of reference	Publication
		Floristic	Durand & Schinz 1895
Senegal	Basse Casamance	Floristic	Berhaut 1988
Seychelles		Floristic	Durand & Schinz 1895
Sierra Leone		Floristic	Durand & Schinz 1895
			Täckholm & Drar 1941
South Africa	Mpumalanga		Marloth 1915
	Eastern Cape		Mengistu et al. 2012
	Limpopo Province		Mengistu et al. 2012
	Western Cape	Brochure	Republic of South Africa 2011
	Gauteng		van Rensburg et al. 2011
Sudan	Kordofan	Floristic	Durand & Schinz 1895
	Kordofan	Floristic	Broun & Massey 1929
	Bahr El Ghazal	Floristic	Broun & Massey 1929
	Southern and Central	Floristic	Andrews 1956
	Bahr el Ghazal		Täckholm & Drar 1941
Sudan/Uganda		Floristic	Friis & Vollesen 1998
Swaziland			Antonsson-Ogle 1984
Tanganyika			Täckholm & Drar 1941
Tanzania			Täckholm & Drar 1941
	Pemba, Zanzibar	Floristic	Williams 1949
	East Africa	Herbarium	Mayo 1985
	Pemba	Ethnographic	Walsh 2009
Uganda	Lake Victoria (known at the time as Ukérévé)	Floristic	Durand & Schinz 1895
			Täckholm & Drar 1941
	East Africa	Herbarium	Mayo 1985
Zimbabwe			Henderson 2007
			Murdock 1959

The identification of non-dietary crop products of *Eleusine coracana* (L.) Gaertn. ssp. *coracana, Pennisetum glaucum* (L.) R. Br., and *Sorghum bicolor* (L.) Moench by phytolith analysis

Welmoed A. Out

Department of Archaeological Science and Conservation
Moesgaard Museum
Moesgaard Allé 15
8270 Højbjerg, Denmark

E-mail: wo@moesgaardmuseum.dk

Marco Madella

CaSEs Research Group, ICREA, Department of Humanities
University Pompeu Fabra & IMF-CSIC
C/Trias Fargas 25–27
08005 Barcelona, Spain

E-mail: marco.madella@icrea.cat

Abstract. Non-dietary cereal crop products, consisting primarily of leaves and culms, regularly remain undetected or unidentified in the archaeobotanical record in many parts of the world because of deposition processes and taphonomic factors. Phytolith analysis may help filling this gap in the record. In this paper, we investigate whether phytolith morphometrics allow for taxonomic identification and plant part detection of the crops *Eleusine coracana* L. Gaertn. (finger millet); *Pennisetum glaucum* (L.) R. Br. (pearl millet); and *Sorghum bicolor* (L.) Moench (sorghum), all millets. Preliminary results indicate extensive possibilities for this approach.

Keywords: cereal identification, crop by-products, millets, phytolith morphometry

1. Introduction

The detection and identification of crop products at agricultural sites are key issues in archaeobotany because they can lead to a better understanding of former agricultural practices. In most depositional environments, botanical remains are preserved only after carbonization. In the case of crop plants, generally only the caryopses and solid chaff parts come into contact with fire during crop processing. If culms and leaves do come into contact with fire, they usually simply burn away; if, by exception, they become carbonized, they are very sensitive to mechanical destruction.

While macroremains and chaff of free-threshing crops are rarely retrieved at archaeological sites in Africa (e.g. Giblin & Fuller 2011; Young 1999), the depositional and taphonomic aspects mentioned above have resulted in an even greater gap in our knowledge about non-dietary crop products consisting of leaves, culms, and chaff at many archaeological sites, despite the economic value (Van der Veen & Tabinor 2007) of these products. Culms, leaves, and chaff can, for example, be used as animal fodder, fuel, construction material, or temper (e.g. Grubben & Partohardjono 1996). Detection and identification of these products at archaeological sites all over the world is important because it will lead to a better understanding of the use of crop plants, the function of archaeological features, and the socio-economic aspects of ancient agricultural societies (Harvey & Fuller 2005). Phytolith analysis allows for the detection and identification of these non-dietary crop products.

Phytoliths are microscopic bodies formed in the intra- and extracellular spaces of tissues of many living plants. They consist of amorphous hydrated opal silica (SiO_2). When plant material decomposes, phytoliths are released into the environment. Because phytoliths are mineral in composition, they resist decomposition, even after burning or ingestion (Piperno 2006). These unique characteristics enable the retrieval of phytoliths from most types of archaeological deposits and contexts all over the world. Because most phytoliths are formed in the intracellular spaces, they have the anatomical characteristics of the original plant cells and tissues. Although not all phytoliths are morphologically unique, their characteristics often allow for the distinction of both plant taxa and plant parts.

The identification of crop plants by phytolith analysis is a robust field of research. Since the 1980s, neotropical crops, such as *Zea mays* L. (corn) and squashes and gourds, have received major attention (Chandler-Ezell et al. 2006; Pearsall et al. 2003; Piperno 2009). Since the early 1990s, identification criteria have been developed for Eurasian cereals, such as *Avena* spp. (oat); *Hordeum* spp. (barley); and *Triticum* spp. (wheat) (Ball et al. 2009; Portillo et al. 2006; Rosen 1992). More recently, criteria have been developed for *Panicum miliaceum* L. (common millet) and *Setaria italica* (L.) P. Beauvois (foxtail millet) (Lu et al. 2009). Depending on the taxon and the amount of research invested, identification to the taxonomic level of family, genus, and species is possible (e.g. Ball et al. 1996).

The first systematic study within the framework of archaeobotany of African wild and domesticated grasses by means of phytoliths was conducted by Fahmy (2008), who studied the bilobates (short cells) from the leaves of 66 Paniceae grasses from the Sahel region of West Africa. Comparison of morphotypes and measurements resulted in the distinction of 25 subgroups. Radomski and Neumann (2011) studied the morphology of inflorescence phytoliths from 18 mostly West African grasses, showing, among other things, that *Sorghum bicolor* (L.) Moench (sorghum) is characterized by unique phytoliths and that it is possible to distinguish groups within Panicoideae, including *Pennisetum glaucum* (L.) R. Br. (pearl millet) on the basis of bilobates and crosses (both short cells). Madella et al. (2013), studying long cell phytoliths from glumes, show that is it possible to distinguish between two groups of millets occurring in Asia and Africa, respectively. Differences between *Sorghum* and *Pennisetum* were also observed, with long cells of the latter showing more rounded cell joints and a scrobiculate surface. Jattisha & Sabu (2012) studied the phytolith composition for a group of Chloridoideae grasses from southern India. However, the proposed identification key cannot be used on phytolith assemblages from excavations because it relies on proportions (%) of phytolith morphologies for the species identification, which are unknown in the case of archaeological assemblages. Despite the high relevance of most of the above-mentioned studies, there is a need for further development of identification criteria for some major African crops based on phytoliths.

Apart from some studies on phytoliths from leaves of *Oryza sativa* L. and *Musa* sp. (rice and wild banana) (Ball et al. 2006; Pearsall et al. 1995; Whang et al. 1998; Zhao et al. 1998) and a single study on those of *Triticum monococcum* L. (einkorn) (Ball et al. 1993), most of the work on the identification of domesticated crops by phytolith analysis has focussed on inflorescences. This is so for two reasons. First, people concentrated on those parts that are anatomically mostly related to the edible part of the crops and that are therefore expected to end up in on-site archaeological deposits. Second, the inflorescences produce phytoliths that are considered to be most diagnostic for taxonomic identification (e.g. the dendritics for cereals; see Piperno [2006: 78]).

What emerges from this short overview is that there is a need to establish phytolith identification criteria for additional crop plants used in various continents of the world, in particular for plant parts other than inflorescences. Therefore, the goal of this research is to be able to identify leaves and culms of the millets *Eleusine coracana* (L.) Gaertn. ssp. *coracana* (finger millet); *Pennisetum glaucum*; and *Sorghum bicolor* by means of phytoliths. Finger millet, pearl millet, and sorghum are major crops in Africa and they are known from archaeological sites in Africa, South Asia, and South-East Asia (Weber & Fuller 2008). The details of their domestication history are still unknown. The earliest finds of domesticated *Eleusine* in Africa date to the first millennium AD (Giblin & Fuller 2011). Domesticated *Pennisetum* is present in Africa from c. 2500 BC onward (Manning et al. 2011); it became widespread in this continent from the first half of the second millennium BC onwards. *Sorghum* was presumably domesticated in Africa before the second millennium BC

(Zohary et al. 2012). These crops were also introduced in India, starting c. 2000 BC (Fuller & Madella 2002).

Our ability to distinguish the input of different types of plant parts based on phytolith analysis of plant parts other than the inflorescence relates to the anatomical characteristics of the tissues at the cellular level (e.g. Madella 2007). Studies by Ball et al. (1992) and Madella et al. (2009) show that there is also an environmental factor affecting phytolith production (in terms of both shape/size and quantity). This environmental factor can, however, be taken into account. Therefore, a major question to resolve is whether the dimensional variation among phytolith morphotypes of different taxa, as caused by genes and environment, is significant with respect to the variation within plant parts, within plant populations, and within taxa (genus and/or species). The main objective of our ongoing study is, hence, the taxonomic identification of plant parts from some major millet crops by phytolith morphometrics (analysis of size and shape) of short and long cell phytoliths, as well as assemblage composition/variability. This paper gives an overview of the preliminary results and focuses on short cell phytoliths from leaves, since they show optimal silicification in the studied modern material. Quantitative statements are based on general observations and comparison of all tissue samples.

2. Materials and methods

The plant material discussed in this work consists of two leaf samples and two culm samples from two plants from two different populations each of finger millet, pearl millet, and sorghum (see Table 1), which were grown at the Institució Milà i Fontanals in Barcelona in 2011–2012. Caryopses of these populations were kindly provided by the Plant Genetic Resources Conservation Unit, the North Central Regional Plant Introduction Station, and the National Small Grains Germplasm Research Facility, all part of the National Plant Germplasm System, Agricultural Research Service of the United States Department of Agriculture.

In order to prepare tissue fragments to study the phytoliths *in situ*, culm and leaf fragments were collected at the end of the natural life cycle, when the plants were completely dead. The plant tissues were washed for 15 minutes in an ultrasonic bath to remove possible dust particles and then placed in household bleach (<5% sodium hypochlorite) for c. 5 hours to remove part of the organic matter. Subsequently, the samples were soaked in demineralized water overnight to eliminate the bleach. Then, the tissues were dehydrated by ethanol (96%) replacement. Finally, the tissues were mounted on microscopy slides using Entellan. Occasionally, samples were mounted in demineralized water. Observations and photographs were carried out at a magnification of 630×. This procedure allows for the two-dimensional study of size and shape aspects. In this paper we discuss the phytoliths observed on the leaf veins, to a maximum of 10 phytoliths per vein to assure random sampling within individual leaves. Morphotype frequencies were estimated by non-quantitative analysis on all available samples, totalling 72. Morphometric analysis was carried out using a software tool developed in the open-source program FIJI (Schindelin et al. 2012) and applying internationally recognized standard measurements of shape and size (Out et al. 2014). Data presented here concern the parameters Shape and RFactor, defined as:

Shape = Perimeter²/Area

RFactor = Convex Hull /(Feret*π)

where Convex Hull is the length of a string tightly drawn around the object and Feret is the largest axis length of the measured object.

Table 1. Experimental set-up of the plant material described in this study.

Subfamily	Genus	Species	Populations	Plants per population	Plant parts	Samples per plant part
Chloridoideae	Eleusine	coracana	2	2	culm	2
					2 leaf blades	2
Panicoideae	Pennisetum	glaucum	2	2	culm	2
					2 leaf blades	2
Panicoideae	Sorghum	bicolor	2	2	culm	2
					2 leaf blades	2

3. Preliminary results

3.1 Identification of plant parts

Figures 1–3 show tissue fragments of culms, leaf blades, and inflorescences of *Eleusine coracana*, *Pennisetum glaucum*, and *Sorghum bicolor*. Samples mounted in water instead of Entellan exhibited better visibility of phytoliths (see Fig. 2, culm). Comparisons among the various plant parts from the various samples and between plants of two populations grown under equal environmental conditions show that different anatomical parts are characterized by differential silicification and different phytolith morphologies. Concerning differential silicification, the degree of silicification in the culms of all taxa is low in both populations. On the other hand, silicification of the leaf edges is substantially higher, because these are the parts with highest evapo-transpiration (e.g. Madella et al. 2009; Powers-Jones et al. 1998). The low culm silicification is not surprising because water in culms is quickly transferred to leaves and flowers, with little water loss taking place in the organ. These preliminary observations need to be substantiated by further studies with populations from other parts of the world to investigate the variability of silicification under wider-ranging environmental conditions. Studying differential silicification is relevant for archaeology because it affects the interpretation of silica skeletons dominated by smooth long cells that are currently often interpreted as culm/leaf. If culms show little silicification, the recovered skeletons will concern primarily leaves, and culms will be even more underrepresented in the archaeological record than leaves.

The frequency of the morphotypes differs among the plant parts due to anatomical differences. Culms and leaves of the three species under study show considerable numbers of short cells and psilate (smooth) or lightly sinuate long cells. Leaves show the unique presence of bulliform cells, and they have higher proportions of stomata and of interstomatal cells. In contrast to leaves and culms, long cells showing sinuate to dentate to spiny ornamentations dominate the inflorescences. Dendritic cells were rarely observed (but see Radomski and Neumann [2011] for the regular presence of dendritics in *Sorghum*). Stomata, interstomatal cells, and psilate long cells are scarce in inflorescences. The frequency of short cells varies within the inflorescence. Analysis of dry-ashed samples will allow for quantitative assessment of the relative production of phytolith morphotypes among and within organs.

3.2 Taxonomic identification

Short cells and other cell types from a variety of plant species seem to retain sufficient genetic variability to be useful in taxonomical identification (e.g. Dorweiler & Doebley 1997; Mulholland & Rapp 1992; Piperno et al. 2002). Confirming the known differences in phytolith production among the various grass subfamilies (see Twiss et al. 1969), the short cell assemblage of *Eleusine coracana* leaves (Chloridoideae) is characterized by saddles, while that of *Pennisetum glaucum* and *Sorghum bicolor* (Panicoideae) is dominated by bilobates (see Figs. 2 and 3). Both *Pennisetum* and *Sorghum* leaves also displayed notched bilobates and trilobates, while *Pennisetum* produces symmetric and asymmetric polylobates. Since the frequency of polylobates is higher in *Pennisetum*, this may be of importance for taxonomic identification.

Figures 2 and 3 show the variability of bilobate short cells in leaves of *Pennisetum glaucum* and *Sorghum bicolor*. Both species show bilobates with a wide range of shapes, including typical bilobates with a short or long shank and convex, concave, and flattened margins, but also more squarish bilobates that show similarity with crosses (the indentation between the four lobes is, however, not clearly discernible). Our morphometric analysis of bilobates and crosses from the leaves of *Pennisetum glaucum* and *Sorghum bicolor* is ongoing. Preliminary results based on Shape and RFactor show that the bilobates of the two genera do not differ from each other (Fig. 4). While the presented morphometric parameters show less variation between the two *Pennisetum* populations than between the two *Sorghum* populations, other parameters also show some minor variation between the two *Pennisetum* populations.

Long cells—a phytolith morphotype that has not been explored before because it is considered to form under little genetic control (Madella et al. 2009) and that is often morphologically recurring—may be more relevant for taxonomic identification. Initial results from our work seem to highlight variable diagnostic value at the morphological level. Long cells of *Pennisetum* leaves have an assortment of psilate, undulated, and dentate long cells. Tissues of *Eleusine* leaves show not only psilate and undulated long cells, but also a regular (higher) occurrence of spiny or dentate long cell phytoliths (Fig. 1). These observations indicate that spiny and dentate long cells can be more distinctive of *Eleusine* and that they cannot be interpreted uniquely as inflorescence morphotypes (as is currently common practice). The long cells of *Sorghum* leaves often show characteristic undulated sides, particularly at the edge of the leaf (Fig. 3). This may be meaning-

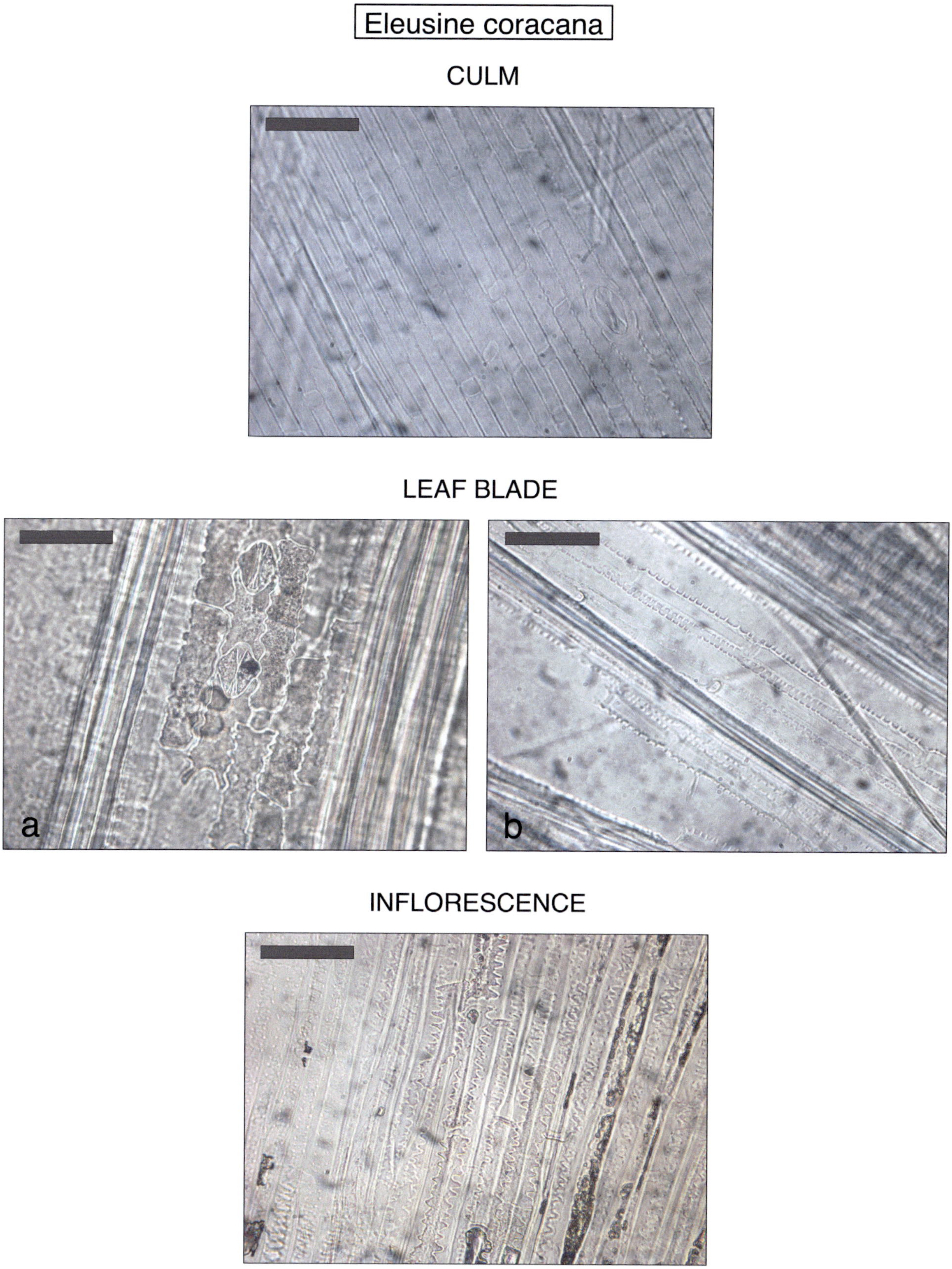

Figure 1. *Eleusine coracana*. Culm: Long and short cells, showing little silicification; Leaf blade, a: veins and predominantly long cells, stomata, and interstomatal cells, b: spiny long cells, not silicified; Inflorescence: lemma, long cells. All photographs: scale bar = 50 μm.

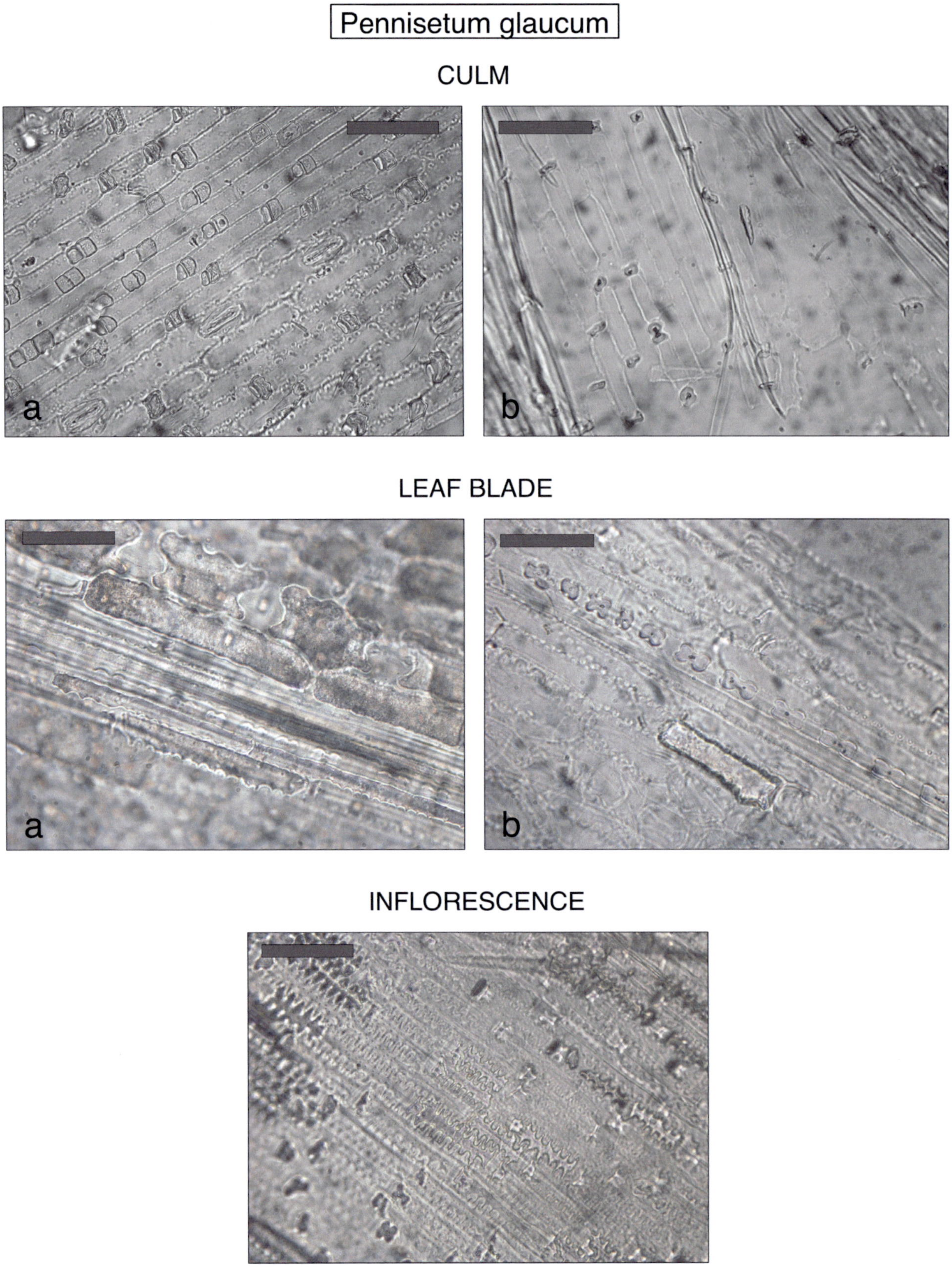

Figure 2. *Pennisetum glaucum*. Culm, a: mounted in water, b: mounted in Entellan; Leaf blade, a: vein, long cells, interstomatal cells and short cells, b: variability of bilobates; Inflorescence: long cells, short cells.

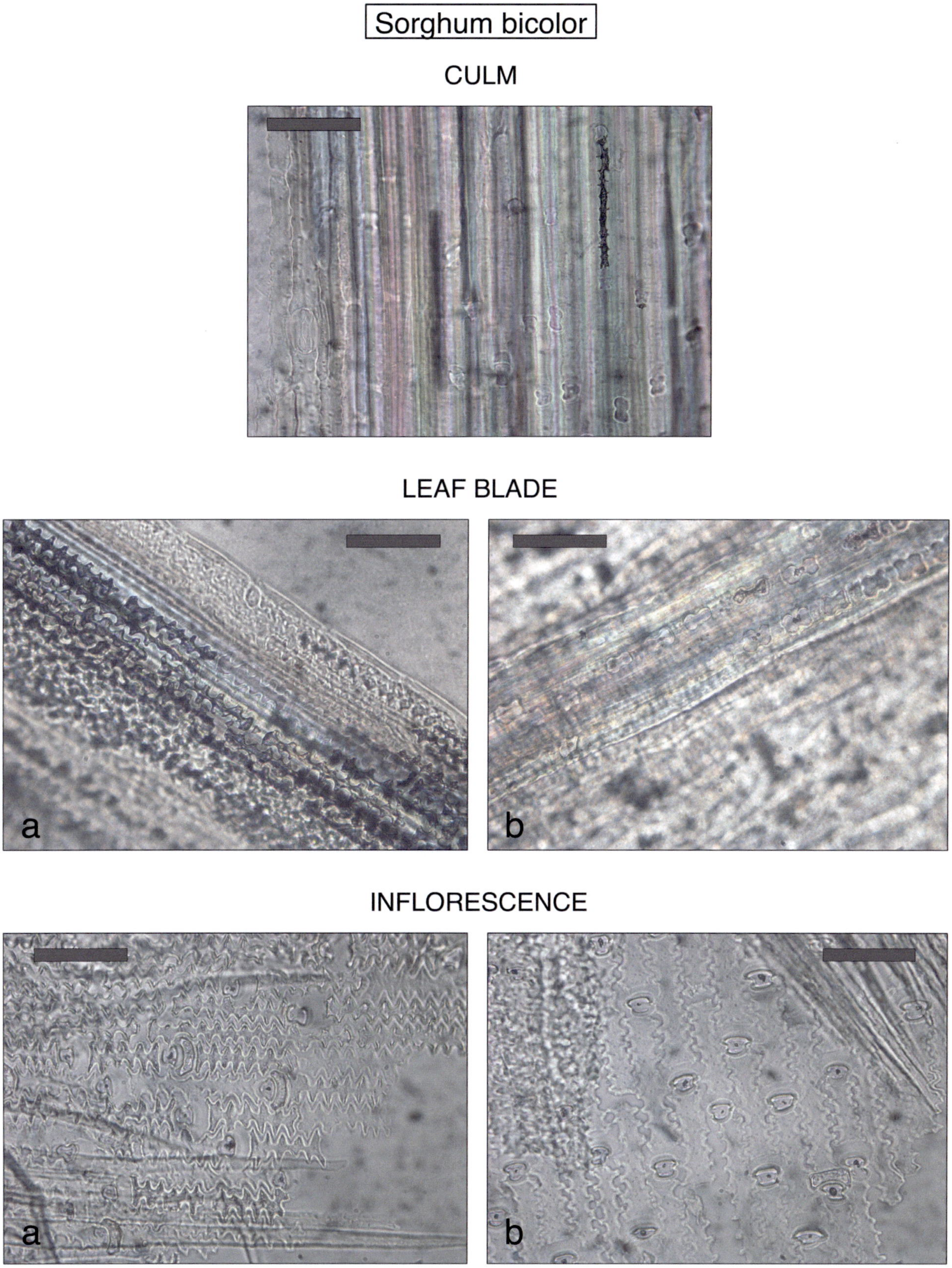

Figure 3. *Sorghum bicolor*. Culm: long cells and short cells; Leaf blade, a: leaf edge showing primarily long cells, b: variability of bilobates; Inflorescence, a and b: glumes: long cells, short cells, and cork cells.

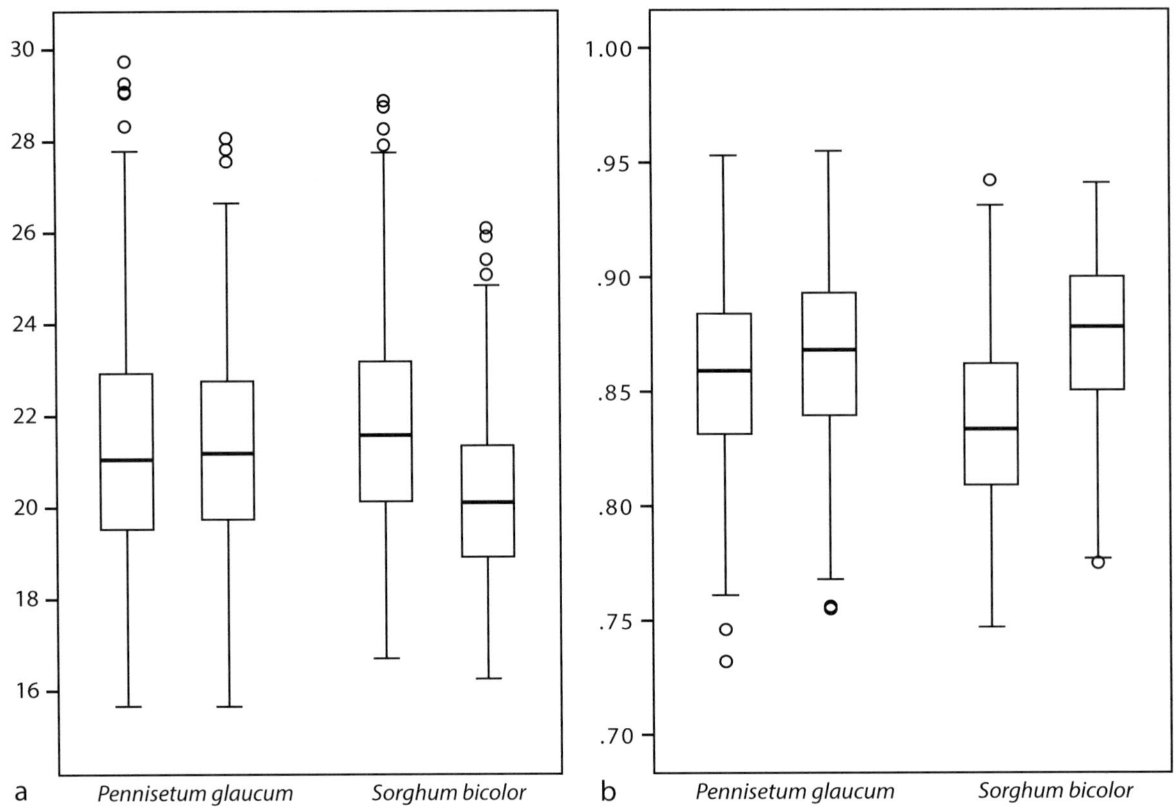

Figure 4. Mean value of measurements of (a) Shape and (b) RFactor (see text) of bilobate phytoliths from the leaves of two populations of *P. glaucum* and *S. bicolor* (N = 400 phytoliths per population). Populations: *P. glaucum* PI 584509 and PI 661490, *S. bicolor* PI 248298 and PI 181080. O = outlier (>1.5 and <3 interquartile range).

ful at a taxonomic level. Future morphometric analysis is required to assess the level of significance and confidence for taxonomic identification of the different long cell phytoliths from leaves.

4. Discussion and conclusions

The results of our experimental approach lead us to the following three observations. First, it is possible to distinguish among plant parts by comparing the phytolith morphotypes of culms, leaf blades, and inflorescences of *Eleusine coracana*, *Pennisetum glaucum*, and *Sorghum bicolor*. Second, from a taxonomic point of view, it is possible to separate some of these species (*Eleusine* versus *Pennisetum/Sorghum*) on the basis of short cell production, because they pertain to two different sub-families (Chloridoideae and Panicoideae). Third, the initial results of the morphometric analysis show that there seems to be little or no potential to distinguish between bilobate phytoliths produced in the leaves of *Pennisetum* versus the ones produced in the leaves of *Sorghum*. For taxonomic identification between these two taxa, an approach based on morphotypes of both short and long cells from leaves (cf. Madella et al. 2013; Radomski & Neumann 2011) might therefore be more promising.

There are some caveats concerning the application of the obtained morphometric results and related identification criteria. Because this study only includes a single species in each genus, the results should be considered valid at the genus level only; they cannot exclude other species within each relevant genus. This means that the outcomes can be applied to contexts in regions and sites where the occurrence of at least one of these genera has been demonstrated by macroremains analysis. Furthermore, the results should not be applied to early agricultural sites until the studied species have been compared with their wild relatives. A first start has been made on these comparisons by, among others, Kumar Tripathi, Mishra et al. (2012); Kumar Tripathi, Kumar Chauhan et al. (2012); and Mercader et al. (2010). In a related issue, the current study also shows the need for the broader development of phytolith systematics and reference collections.

A second important point arising from this research is that phytolith morphometric analysis to identify plant resources can ideally be best applied to closely constrained contexts, where the presence of a limited number of taxa, resulting from a single deposition event, can be expected (cf. Shillito 2012). This suggests that phytolith samples should preferably be collected from closed or well-controlled contexts. In such cases, a combination of frequency analysis (proportions of different phytolith morphologies), morphotype analysis and, when relevant, morphometric analysis can be used to test whether non-dietary crop by-products are present.

Much of the variation in phytolith composition of different plant parts, whether within the same taxon or among different taxa, remains to be understood more clearly. Nevertheless, this line of research has the potential to play a major role in the detection and identification of dietary and non-dietary crop (by-)products at archaeological sites, regardless of the state of preservation of organic material.

Acknowledgements

The paper is dedicated to A. Fahmy (†) because of his outstanding contributions to African archaeobotany and phytolith systematics. The research was funded by the Marie Curie Intra-European Fellowship PHYTORES (273610, 2011–2013). The authors are grateful to the Plant Genetic Resources Conservation Unit, the North Central Regional Plant Introduction Station, and the National Small Grains Germplasm Research Facility of the United States Department of Agriculture for supplying the grass caryopses and to D.Q. Fuller of University College London for kindly providing reference material of *Pennisetum* and *Sorghum* used for the photographs of the inflorescences.

References

Ball T.B. & Brotherson J.D., 1992. The effect of varying environmental conditions on phytolith morphometries in two species of grass (*Bouteloua curtipendula* and *Panicum virgatum*). *Scanning Microscopy* 6 (4), pp. 1163-1181.

Ball T.B., Brotherson J.D. & Gardner J.S., 1993. A typologic and morphometric study of variation in phytoliths from einkorn wheat (*Triticum monococcum*). *Canadian Journal of Botany* 72, pp. 1182-1192.

Ball T.B., Ehlers R. & Standing M.D., 2009. Review of typologic and morphometric analysis of phytoliths produced by wheat and barley. *Breeding Science* 59, pp. 505-512.

Ball T., Gardner J.S. & Brotherson J.D., 1996. Identifying phytoliths produced by the inflorescence bracts of three species of wheat (*Triticum monococcum* L., *T. dicoccon* Schrank., and *T. aestivum* L.) using computer-assisted image and statistical analyses. *Journal of Archaeological Science* 23, pp. 619-632.

Ball T., Vrydaghs L., Van Den Hauwe I., Manwaring J. & De Langhe E., 2006. Differentiating banana phytoliths: Wild and edible *Musa acuminata* and *Musa balbisiana*. *Journal of Archaeological Science* 33, pp. 1228-1236.

Chandler-Ezell K., Pearsall D.M. & Zeidler J.A., 2006. Root and tuber phytoliths and starch grains document manioc (*Manihot esculenta*), arrowroot (*Maranta arundinacea*) and llerén (*Calathea* sp.) at the Real Alto site, Ecuador. *Economic Botany* 60, pp. 103-120.

Dorweiler J.E. & Doebley J., 1997. Developmental analysis of teosinte glume architecture: A key locus in the evolution of maize (Poaceae). *American Journal of Botany* 84, pp. 1313-1322.

Fahmy A.G., 2008. Diversity of lobate phytoliths in grass leaves from the Sahel region, West Tropical Africa: Tribe Paniceae. *Plant Systematics and Evolution* 270, pp. 1-23.

Fuller D. & Madella M., 2002. Issues in Harappan archaeobotany: Retrospect and prospect. In: Settar S. & Korisettar R. (eds.), *Indian Archaeology in Retrospect.* Vol. II. Protohistory. Publications of the Indian Council for Historical Research, New Delhi, pp. 317-390.

Giblin J. & Fuller D., 2011. First and second millennium A.D. agriculture in Rwanda: Archaeobotanical finds and radiocarbon dates from seven sites. *Vegetation History and Archaeobotany* 20 (4), pp. 253-265.

Grubben G.J.H. & Partohardjono S., 1996. *Plant Resources of South-East Asia.* Vol. 10. Cereals. Prosea Foundation, Bogor, Indonesia.

Harvey E.L. & Fuller D.Q., 2005. Investigating crop processing using phytoliths analysis: The example of rice and millets. *Journal of Archaeological Science* 32, pp. 739-752.

Jattisha P.I. & Sabu M., 2012. Phytoliths as a tool for the identification of some Chloridoideae grasses in Kerala. International Scholarly Research Network Botany 2012, article ID 246057, doi:10.5402/2012/246057.

Kumar Tripathi D., Mishra S., Kumar Chauhan D., Prakash Tiwari S. & Kumar C., 2012. Typological and frequency based study of opaline silica (phytolith) deposition in two common Indian *Sorghum* L. species. *Proceedings of the Indian National Science Academy Part B Biological* Science 83 (1), pp. 97-104.

Kumar Tripathi D., Kumar Chauhan D., Kumar D. & Prakash Tiwari S., 2012. Morphology, diversity and frequency based exploration of phytoliths in *Pennisetum typhoides* Rich. *National Academy Science Letters India* 35 (4), pp. 285-289.

Lu H., Zhang J., Wu N., Liu K.-B. & Xu D., 2009. Phytolith analysis for the discrimination of foxtail millet (*Setaria italica*) and common millet (*Panicum miliaceum*). *Plos One* 4 (2), e4448. doi 10.1371/journal.pone.0004448.

Madella M., 2007. The silica skeletons from the anthropic deposits. In: Whittle A. (ed.), *The Early Neolithic on the Great Hungarian Plain: Investigations of the Körös Culture Site of Ecsegfalva 23, County Békés*. Varia Archaeologica Hungarica XXI. Archaeological Institute of the Hungarian Academy of Sciences, Budapest, pp. 447-460.

Madella M., Jones M.K., Echlin P., Powers-Jones A. & Moore M., 2009. Plant water availability and analytical microscopy of phytoliths: Implications for ancient irrigation in arid zones. *Quaternary International* 193, pp. 32-40.

Madella M., Lancelotti C., García-Granero J.J., 2013. Millet microremains—an alternative approach to understand cultivation and use of critical crops in Prehistory. *Archaeological and Anthropological Sciences*, doi 10.1007/s12520-013-0130-y.

Manning K., Pelling R., Higham T., Schwenniger J.-L. & Fuller D.Q., 2011. 4500-year old domesticated pearl millet (*Pennisetum glaucum*) from the Tilemsi valley, Mali: New insights into an alternative domestication pathway. *Journal of Archaeological Science* 38 (2), pp. 312-322.

Mercader J., Astudillo F., Barkworth M., Bennett T., Esselmont C., Kinyanjui R., Laskin Grossman D., Simpson S. & Walde D., 2010. Poaceae phytoliths from the Niassa Rift, Mozambique. *Journal of Archaeological Science* 37, pp. 1953-1967.

Mulholland S. & Rapp G., 1992. A morphological classification of grass silica bodies. In: Rapp G. & Mulholland S. (eds.), *Phytolith Systematics: Emerging Issues*. Plenum Press, New York, pp. 65-89.

Out W.A., Pertusa Grau J.F. & Madella M., 2014. A new method for morphometric analysis of opal phytoliths from plants. *Microscopy and Microanalysis* 20(6), pp. 1876-1887.

Pearsall D.M., Chandler-Ezell K. & Chandler-Ezell A., 2003. Identifying maize in neotropical sediments and soils using cob phytoliths. *Journal of Archaeological Science* 30, pp. 611-627.

Pearsall D.M., Piperno D.R., Dinan E.H., Umlauf M., Zhao Z. & Benfer R.A., 1995. Distinguishing rice (*Oryza sativa* Poaceae) from wild *Oryza* species through phytolith analysis: Results of preliminary research. *Economic Botany* 49 (2), pp. 183-196.

Piperno D.R., 2006. *Phytoliths: A Comprehensive Guide for Archaeologists and Paleoecologists*. Altamira Press, Lanham.

Piperno D.R., 2009. Identifying crop plants with phytoliths (and starch grains) in Central and South America: A review and an update of the evidence. *Quaternary International* 193, pp. 146-159.

Piperno D.R., Holst I., Wessel-Beaver L. & Andre T.A., 2002. Evidence for the control of phytolith formation in *Cucurbita* fruits by the hard rind (Hr) genetic locus: Archaeological and ecological implications. *Proceedings of the National Academy of Sciences* 99 (16), pp. 10923-10928.

Portillo M., Ball T. & Manwaring J., 2006. Morphometric analysis of inflorescence phytoliths produced by *Avena sativa* L. and *Avena strigosa* Schreb. *Economic Botany* 60 (2), pp. 121-129.

Powers-Jones A.H., Echlin P. & Jones M.K., 1998. Investigating ancient plant–water relations by the means of analytical microscopy of phytoliths. In: Pinilla A., Tresserras J.J. & Machado M.J. (eds.), *Estado actual de los estudios de fitolitos en suelos y plantas*. [Proceedings of the first European meeting on phytolith research.] Centro de Ciencias Medioambientales, Madrid, pp. 245-253.

Radomski K.U. & Neumann K., 2011. Grasses and grinding stones: Inflorescence phytoliths from modern West African Poaceae and archaeological stone artefacts. In: Fahmy A.G., Kahlheber S. & D'Andrea A.C. (eds.), *Windows on the African Past: Current Approaches to African Archaeobotany*. [Proceedings of the sixth international workshop for African Archaeobotany.] Africa Magna Verlag, Frankfurt, pp. 153-166.

Rosen A., 1992. Preliminary identification of silica skeletons from Near Eastern archaeological sites: An anatomical approach. In: Rapp G. & Mulholland S. (eds.) *Phytolith Systematics: Emerging Issues*. Plenum Press, New York, pp. 129-147.

Schindelin J., Arganda-Carreras I., Frise E., Kaynig V., Longair M., Pietzsch T., Preibisch S., Rueden C., Saalfeld S., Schmid B., Tinevez J.-Y., White J.D., Hartenstein V., Eliceiri V., Tomancak P. & Cardona A., 2012. Fiji: An open-source platform for biological image-analysis. *Nature Methods* 9 (7), 676-682.

Shillito L.-M., 2012. Grains of truth or transparent blindfolds? A review of current debates in archaeological phytolith analysis. *Vegetation History and Archaeobotany* 22 (1), 71-82.

Twiss P.C., Suess E. & Smith R.M., 1969. Morphological classification of grass phytoliths. *Soil Society of America Proceedings* 33 (1), pp. 109-115.

Van der Veen M. & Tabinor H., 2007. Food, fodder and fuel at Mons Porphyrites: The botanical evidence. In: Maxfield V.A. & Peacock D.P.S. (eds.), *Survey and Excavation at Mons Porphyrites 1994–1998*. Vol. 2. The excavations. Egypt Exploration Society, London, pp. 83-142.

Weber S.A. & Fuller D.Q., 2008. Millets and their role in early agriculture. *Pragdhara* 18, pp. 69–90.

Whang S.S., Kim K. & Hess W.M., 1998. Variation of silica bodies in leaf epidermal long cells within and among seventeen species of *Oryza* (Poaceae). *American Journal of Botany* 85 (4), pp. 461-466.

Young R., 1999. Finger millet processing in East Africa. *Vegetation History and Archaeobotany* 8, pp. 31-34.

Zhao Z.J., Pearsall D.M., Benfer Jr. R.A. & Piperno D.R., 1998. Distinguishing rice (*Oryza sativa* Poaceae) from wild *Oryza* species through phytolith analysis, II: Finalized method. *Economic Botany* 52, pp. 134-145.

Zohary D., Hopf M. & Weiss E., 2012. Domestication of plants in the Old World. 4th ed. Oxford University Press, Oxford.

Charred plant macroremains (seeds, fruits) and phytoliths from villa E12.10 at Amara West, a pharaonic town in northern Sudan

Philippa Ryan

Department of Conservation and Scientific Research
British Museum
Great Russell Street
London WC1B 3DG, United Kingdom

E-mail: pryan@thebritishmuseum.ac.uk

Caroline R. Cartwright

Department of Conservation and Scientific Research
British Museum
Great Russell Street
London WC1B 3DG, United Kingdom

E-mail: ccartwright@thebritishmuseum.ac.uk

Neal Spencer

Department of Ancient Egypt and Sudan
British Museum
Great Russell Street
London WC1B 3DG, United Kingdom

E-mail: nspencer@thebritishmuseum.ac.uk

Abstract. This paper presents macrobotanical (seeds, fruits) and phytolith data from an elite house in a pharaonic town in northern Sudan. Amara West, occupied between 1300 and 1000 BC, was founded as the administrative centre of occupied Upper Nubia. Villa E12.10 was situated in an extramural suburb and featured well-preserved food processing installations with associated charred macrobotanical remains. A range of archaeobotanical datasets provide complementary categories of information about plant use in different areas of the villa.

Keywords: charred macroremains, phytoliths, Sudan, Upper Nubia, New Kingdom

1. Introduction

1.1 Amara West

The New Kingdom (c. 1550–1069 BC) control of Upper Nubia, upstream of the second Nile Cataract, was characterized by the foundation of walled towns, provided with temples, administrative buildings, and housing areas, including those at Dokki Gel, Tombos, Kerma, Sesebi, Soleb, and Sai (Edwards 2004; Smith 2003). Under the reigns of Seti I and Ramses II (c. 1294–1213 BC), considerable investment was made in founding further towns, notably those of Aksha and Amara West, while large temples were also constructed in Lower Nubia, most famously that of Abu Simbel.

Amara West, located between the second and third cataracts on the north bank of the Nile as it flows east after Sai Island (Fig. 1), was the administrative centre of Upper Nubia, known as Kush, during the thirteenth and twelfth centuries BC. Excavations by the Egyptian Exploration Society (Spencer 1997, 2002) in 1938–39 and 1947–50 revealed a decorated sandstone temple and a formal residence with *in situ* inscriptions naming holders of the title 'Deputy of Kush', the highest administrative position in the region.

Since 2008, a British Museum research project (www.britishmuseum.org/AmaraWest) has been investigating the nature of ancient lived experience at the site, through excavating residential areas and two contemporary cemeteries. The excellent preservation of houses, storage

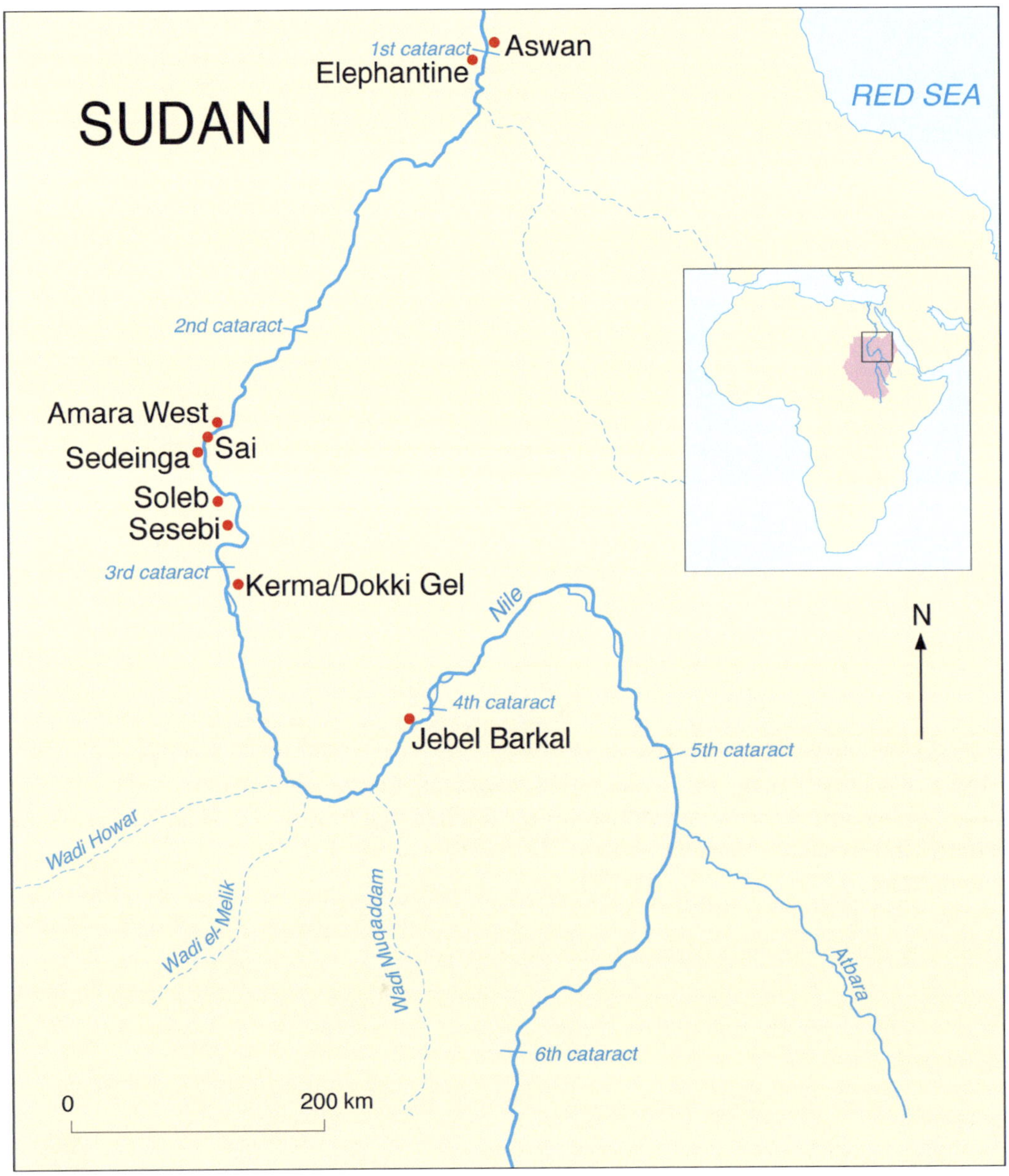

Figure 1. Map showing location of Amara West. (Map: Claire Thorne.)

magazines, and other buildings, both inside the walled town and beyond, allows a detailed consideration of how neighbourhoods developed across time, and of the role of individual, household, and external agencies in shaping this built environment (Spencer 2014, 2015). While the architecture and artefact assemblages are consistent with those found in Egypt proper, a circular building (E12.11) represents a rupture with pharaonic architecture, instead reflecting long-standing Nubian traditions (Spencer 2010). Pottery is mainly Egyptian in style and technology (Spataro et al. 2014), but handmade Nubian cooking pots are one of the most commonly found vessel types, and there is a notable increase in the proportion of such vessels during the twelfth century BC. Through the generosity of the National Corporation of Antiquities & Museums, it is possible to export archaeological samples to undertake analyses in laboratories at the British Museum and collaborating universities. This includes archaeobotanical material that is being used to study plants used for food, fuel, and craft activities (Ryan et al. 2012).

The town was originally founded on an island adjacent to the north bank of the Nile. The failure of a channel north of the town has been dated to the late second millennium BC; this failure may have been one of the primary factors that led to the abandonment of the settlement (Spencer et al. 2012). Land suitable for agricultural exploitation would have been reduced, and accumulations of windblown sand in the settlement would have become increasingly problematic. Nonetheless, the cemeteries were still being used for burials into the ninth century BC (Binder 2011; Binder et al. 2011), long after pharaonic Egypt had lost control of the region. Since the eighth century BC, and continuing until the present day, all major settlements are on the opposite river bank.

This paper presents a detailed comparison of the charred macroremains (seeds, fruits) and phytolith data from a villa (E12.10) outside the town walls.

1.2 Villa E12.10

An area of extramural buildings outside the sandstone west gate of the town was identified through magnetometry survey in 2008 (Spencer & Hay 2012). Several structures could be identified as villas on the basis of similarities with the plans of elite houses at such sites as Tell el-Amarna. At the time of writing, one villa has been fully excavated (E12.10) (Spencer 2009), while excavation has commenced in a villa to the south (D12.5). Preservation across villa E12.10 is uneven. Nearly 2 m of architecture is preserved in Room 13, but the brickwork is heavily truncated at the northern end through wind erosion. Sondages through the floors of three rooms revealed that the villa is founded upon a layer of rubbish characterized by 19th Dynasty pottery; thus this building, and perhaps other large houses in the extramural area, represents a later development in the town layout. Villa E12.10 follows the broadly tripartite division of elite Egyptian houses (Spence 2004), as shown in Figure 2, with a front area comprising an entrance porch (13), courtyard (7), and food processing suite (9–10), followed by a middle area with a broad room (5) set around a hearth, and stairs (4) to the roof or upper storey. The most private inner zone comprises a square room (2) with a low bench against the back wall, as well as two side rooms, including one with a bed alcove (3). The rear part of the house (rooms 1, 2, 3, 5) is characterized by the use of brick pavements underneath the clay floors; rooms 2 and 8 are further distinguished by the application of whitewash to the mud plaster that coats the walls.

Considerable provision was made for food storage and processing facilities. The courtyard, almost certainly unroofed, was provided with five storage bins along its western side. Raised up on schist slabs (presumably to reduce the threat from vermin), these cuboid bins were coated in mud plaster and provided with schist bases sealed with white plaster (Fig. 3). Immediately to the south of these bins was a doorway providing access to two rooms. Room 9 housed four side-by-side emplacements (Feature 2012) for quernstones, constructed from brick, stone, and mud plaster. These provided a setting for individuals to grind cereals (and perhaps other material), squatting over a quernstone, with the resulting products falling into a plastered basin along the south side of the room. The room to the south (10) was provided with seven cylindrical clay ovens with flue holes near their bases (Figs. 3 and 4). These are broadly contemporary, suggesting considerable capacity for bread production.

At nearly 400 m^2 in area, the ground floor of this house falls within the top 10 per cent of house sizes at Tell el-Amarna (Tietze 1985). However, given the remote location of Amara West, controls on display of wealth and prestige through house architecture (whether explicit or implicit) may have been less restrictive than at a royal residence city. The principal expense in constructing such a dwelling was time and labour, as materials (mud, sandstone, schist) were all locally available. Furthermore, the artefact assemblage from the villa was consistent with that found in much smaller dwellings (of 70 to 150 m^2) inside the town wall, and the villa showed no noticeable increase in the number of luxury items, such as calcite, faience, or copper alloy objects.

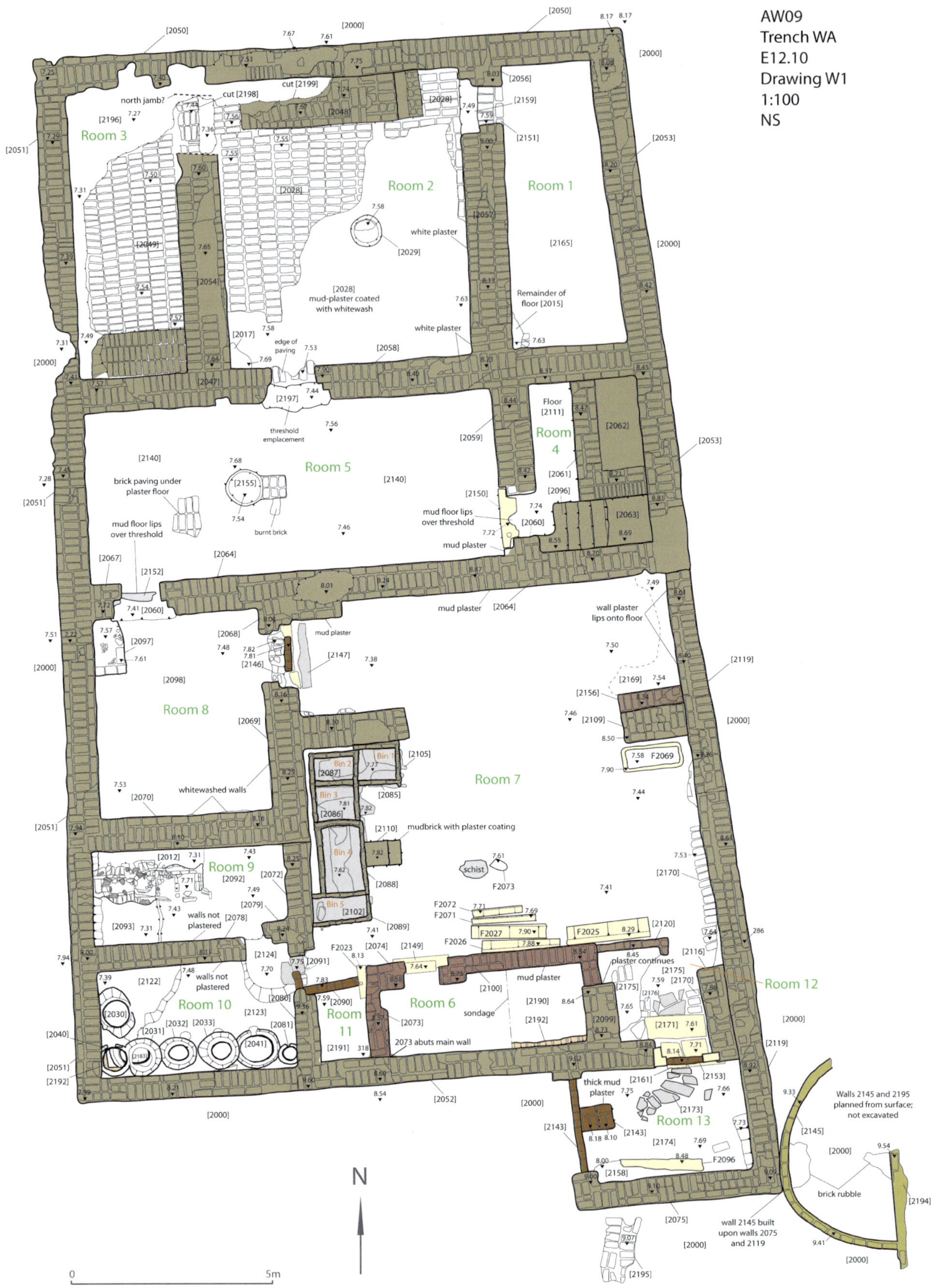

Figure 2. Plan of villa E12.10. (© Trustees of The British Museum.)

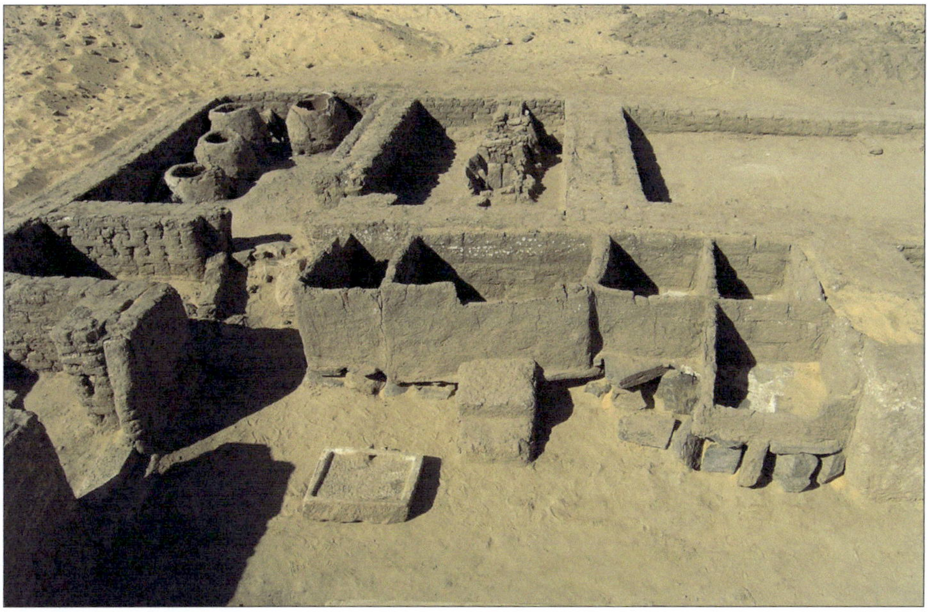

Figure 3. Villa E12.10 showing storage bins in Room 7 (foreground), the grinding emplacement in Room 9 (back, central) and ovens in Room 10 (back, left). (© Trustees of The British Museum.)

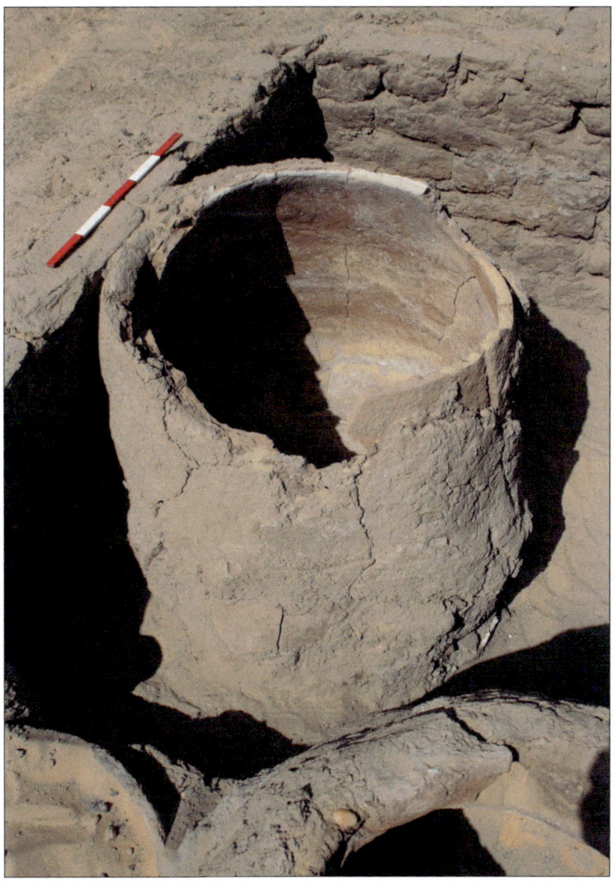

Figure 4. Oven from Room 10. (© Trustees of The British Museum.)

2. Methods

2.1 Sampling rationale

Phytoliths and charred macroremains can provide complementary evidence, and combining analyses is useful for cross-checking trends. Each dataset preserves information about different suites of plant parts. The macrobotanical remains from Amara West have been preserved through charring. Charred seeds can most commonly be used to identify a wider range of taxa than phytoliths, but the latter preserve information about less robust plant parts that do not frequently survive charring, such as leaves and stems. The phytolith record can also provide evidence for plants where charred material has been entirely turned to ash. Phytoliths are formed when monosilicic acid (H_4SiO_4), present in groundwater, is absorbed by plants and deposited as opaline silica, most commonly within or between epidermal plant cells (Piperno 2006). They are abundantly formed in certain monocotyledonous plants, especially grasses, sedges, and palms. They are also formed within various groups of dicotyledonous plants. Phytoliths are released into sediments through both burning and decay. Sediment samples were taken by the excavators during the 2009 field season, with a focus on the well-preserved charred deposits in the southern areas of the building. The sediment samples were processed by a combination of dry sieving and flotation to retrieve the charred plant remains during the 2011 field season and exported to the British Museum for identifi-

Table 1. Context descriptions.

Context	Archaeological sample (AS) number	Room	Description
2022	AS15	E12.10.9	Deposit in easternmost part of grinding table 2012
2025	AS16	E12.10.9	Fill from part of grinding table 2012
2026	AS17	E12.10.9	Fill from part of grinding table 2012
2083	AS 18	E12.10.9	Fill from part of grinding table 2012
2084	AS 18b	E12.10.9	Fill from part of grinding table 2012
2092	AS19	E12.10.9	Floor in SW corner by basin 2093
2016	AS3	E12.10.9	Ash layer, possible location of a fire, against the north wall
2128	AS26	E12.10.10	Charcoal and sandy layer in oven 2035
2182	AS42	E12.10.10	Ash layer in oven 2040
2130	AS28	E12.10.10	Ash deposit in oven 2031
2104	AS23	E12.10.10	Ashy layer in oven 2081
2042	AS 22	E12.10.10	Ash layer in SW corner
2113	AS 24	E12.10.11	Ash and sand layer
2044	AS 9	E12.10.7	Sediments from storage bin
2155	AS 275	E12.10.5	Fill of hearth 2155
2098a	AS 274	E12.10.8	Ash on floor 2098
2098b	S 276	E12.10.8	Floor 2098

cation; further sediment samples were exported for phytolith analysis at the British Museum. Descriptions of the sample contexts are provided in Table 1.

2.2 Charred seed sampling and analysis procedures

The charred seeds were recovered through a combination of dry sieving and flotation (discussed below). The average sample size processed was between 2 and 4 litres. Simple bucket flotation was used due to limited water availability (water was available for 1 to 2 hours a day from one shared tap).

The charred samples were analysed at the British Museum using a low-power binocular microscope, Leica MZ APO (×10 to ×80). A variable pressure scanning electron microscope (VP-SEM), Hitachi S-3700N, was used to further examine selected ancient and modern specimens. The samples were mounted on a scanning electron microscope (SEM) stub using Leit-C Plast carbon cement (a proprietary brand of conductive material with low outgassing properties that is suitable for SEM use). They were examined, uncoated, using the VP-SEM to observe the fine detail of the crucial diagnostic anatomical features. An accelerating voltage of 15 kV was used on most occasions, but sometimes this was raised to 20 kV or lowered to 12 kV, depending on the condition of the sample. For optimal visualization of diagnostic cellular detail, the working distance varied from 23.5 mm to 11.3 mm, as dictated by the individual sample being examined. To eliminate surface charging on non-conducting samples, the chamber pressure (whose unit of measurement is indicated by Pa, being the abbreviation for Pascal) also varied according to the state of preservation of each sample. The highly sensitive, five-segment BSE detector on the VP-SEM enabled detailed interrogation and imaging of the topography of the samples from different orientations and, where necessary, using low accelerating voltages. The 3D mode, rather than the Compositional mode, produced maximum surface topography information.

Identifications of taxa were made using modern reference material as well as reference works (Andrews 1956; Boulos 1999, 2000, 2002, 2005; Braun & Burgstaller 1991; Täckholm 1974). Minimum numbers of individuals (MNI) were recorded for grains and chaff. Spikelet forks and glume bases were counted as individual units, but for the purposes of calculations the results were combined into a total glume base figure (with each spikelet fork equating to two glume bases). MNI were also recorded for wild taxa where possible, with two exceptions. Several whole *Acacia* seeds were present. Fragmented *Acacia* seeds were easily identified, but as it was not possible to calculate MNI, these instances were recorded as 'x' to illustrate presence. Occasionally, whole figs survived in charred form; examples of fruit flesh were recorded as 'x', indicating presence.

2.3 Charred seed preservation in relation to sample processing method

After a period of experimentation it was found that greater numbers of items survived via dry sieving than via flotation (Table 2) and that this observation was particu-

Table 2. Summarized comparison of floated and dry sieved villa samples from processing experiments.

	2083		2025		2026	
	Dry sieve	Flot	Dry sieve	Flot	Dry sieve	Flot
Wheat grain	1	0	6	0	2	
Spikelet fork	80	7	23	4	14	1
Glume base	68	9	68	28	32	6
Barley grain	6	0	2	0	16	2
Barley rachis	33	3	16	0	16	4
Total items	***449***	***209***	***403***	***244***	***384***	***100***
Litres processed	**2**	**2**	**2.5**	**2.5**	**2**	**2**

larly applicable to cereals (found in sieve sizes > 0.5 mm). Spikelet forks also survived more frequently in the dry sieved samples (Fig. 5), whereas glume bases were more common in floated samples. The destruction of charred macroremains in water may relate to aridity and also perhaps to salt crystal dissolution. SEM images of glume bases from flotation samples clearly show disintegration (Figs. 6a and 6b). Such disintegration has previously been recorded at Wadi Kubbaniya by Hillman, Madeyska & Hather (1989).

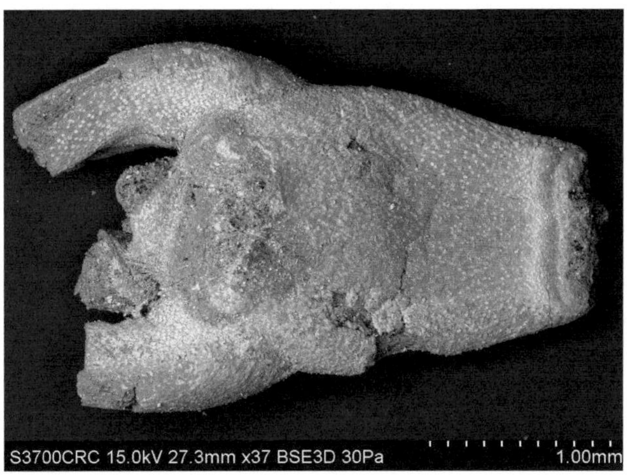

Figure 5. Backscattered electron (BSE) image of an emmer wheat (*Triticum dicoccum*) spikelet fork from a dry sieved sample (oven context 2130). (Image: C.R. Cartwright; © Trustees of The British Museum.)

In contrast, flotation was still the best option for the smallest size fraction (0.3–< 0.5 mm). The grain size distribution of the sediments, which often contained a high

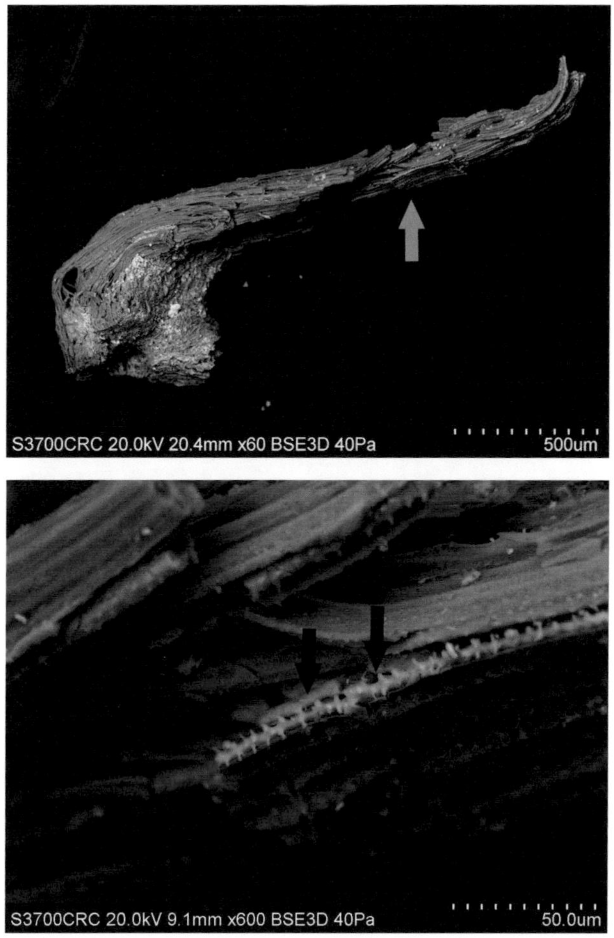

Figure 6. (a) BSE image of an emmer wheat (*Triticum dicoccum*) glume base from flotation sample; (b) detail showing dendritic phytoliths. (Images: C.R. Cartwright; © Trustees of The British Museum.)

proportion of silts and small sands, meant that for many samples, large quantities of sediment became trapped in the 0.3 mm sieve. This makes it difficult to retrieve seeds, as large volumes of small sand particles need sorting. This sorting is also time consuming, allowing only small portions to be sorted in comparison with the flots (for the same amount of time invested). In addition, identification of charred items can be hampered by dirt. In some instances more items per sample were derived from dry-sieved samples than from the equivalent flot samples, but in other instances the reverse was true. This difference relates most probably to the sedimentary composition of the original sample; it is easier to extract seeds from the 0.3–0.5 mm dry fractions of samples dominated by charred items compared with more sandy or very ashy samples. Furthermore, a greater diversity of taxa was retrieved from the flots (0.3–0.5 mm) because it was possible to sort through a larger proportion or even all of the sample, due to the time constraints involved in sorting through the dry sieve 0.3 mm fraction.

2.4 Phytolith sampling and analysis procedures

Phytoliths were extracted from sediments through a series of laboratory steps to remove carbonates, organics, clays, and other non-siliceous elements following methods outlined by Rosen (2005). Phytoliths were mounted onto slides and analysed using a light-transmitting microscope (Olympus BH-2). Typically 300 morphotypes were counted per slide at ×400 magnification.

3 Results

3.1 Macroremains

The charred seeds and other items present are detailed in Table 3. Results are shown for each archaeological context. The table includes original sample volumes, total actual numbers of items, and total numbers of items per litre for each sample. Since samples were dry sieved, as an alternative to showing flot volumes, total charred content was estimated using comparison charts for visual estimation of percentage composition (Bullock et al. 1985; Terry & Chilinger 1955), a geological method sometimes employed in geoarchaeology (Ryan et al. 2011). Actual counts were converted to numbers per litre of sediment, and the frequency (ubiquity) for each taxon was also calculated.

Cereal grains and crop processing by-products were the most frequently represented botanical remains in the charred assemblages. The cereals present were identified as *Triticum dicoccum* Schrank (emmer wheat) and *Hordeum vulgare* subsp. *vulgare* L. (six-row hulled barley), the latter identifiable from well-preserved rachis internodes as well as the ratio of twisted to straight grains. There were high proportions of cereal crop processing by-products in relation to grains and high proportions of barley grains in comparison to wheat grains.

Other food plants present in very small quantities included *Lens culinaris* Medik. (lentil) and *Cucumis* sp. (a genus that includes melons). The taxon with the largest number of seeds, and which was present in the greatest frequency among samples analysed, was *Ficus sycomorus* L. (sycomore fig). High numbers of fig 'seeds' (drupelets) almost certainly reflect the large numbers within individual figs. Whole fig fruit (syconium) fragments were also present. Wild plants, of which many were too charred for identification, include *Acacia* sp., *Glinus lotoides* L. (Molluginaceae family), Trifoliaea (a Fabaceae tribe that includes clover), and the Portulacaceae family. Cyperaceae (sedges) identified were *Fimbristylis* sp. (fimbry) and *Cyperus* sp. (flatsedge). The relatively high numbers of *Fimbristylis* most probably reflect the high level of silicification of these nutlets.

3.1.1 Macroremain results by room

Room 9 (grain processing room): Cereal grains and chaff were abundantly present in samples analysed from different areas of the grain processing structure (Feature 2012), but were absent from floor sample 2092 and present in low amounts from the ashy floor deposit 2016. Samples from the emplacements for querns across Feature 2012 were dominated by chaff, with greater proportions from wheat than from barley. Barley and wheat grains were present in each, and very small numbers of wild grass caryopses were present in two of the three samples. Other seeds present in Room 9 included *Ziziphus spina-christi* (L.) Willd. (Christ's thorn) and *Lens culinaris*. The three samples from Feature 2012 also contained *Ficus sycomorus* 'seeds' and various other categories of small seeds, including sedges.

Room 10 (oven room): All of these samples were from ashy deposits within or near ovens. The cereal assemblages from rooms 9 and 10 were similar, with some possible slight differences (Fig. 7). Wheat grains were only present within one oven sample, and overall there were higher proportions of barley chaff in Room 10 in comparison to Room 9. There were also greater numbers of small wild grass caryopses in the Room 10 samples. Samples also contained other categories of small seeds, including sedges and *Ficus sycomorus*.

Room 11 (small room adjacent to the oven room): The sample, although it was taken from a sandy/ashy deposit, contained no charred seeds.

Room 7 (courtyard): The storage bin sediment sample analysed contained no charred seeds.

Rooms 5 and 8: The two floor samples, one of which was ashy, contained no charred seeds, and hearth deposit 2155 contained only very small numbers of seeds.

3.2 Phytoliths

Table 4 shows the numbers per gram sediment of the different types of phytoliths present. For most of the histograms discussed below, relative percentages were calculated for different morphologies. This is because numbers of phytoliths per gram sediment are affected by the variable sediment components of different context

Table 3. Charred seed counts per litre of sediment. x = present.

Archaeological sample no.	**16**	**17**		**19**	**3**	**22**	**23**	**26**	**28**	**24**	**9**	**274**	**276**	**275**	
Context no.	2025	2026	2083	2092	2016	2042	2104	2128	2130	2113	2044	2098a	2098b	2155	
Litres processed	2.5	2	2	1.5	1	4	3	4	4	3	2	1	3	3	
Proportion charred content	12%	7%	20%	2%	12%	40%	35%	45%	30%	2%	0%	1%	1%	5%	
Total items per litre	126	86	137.5		33	207.3	580.7	155.8	198.3					8.6	
Total items in sample	316	172	275		33	829	1742	623	793					26	
	n per litre sediment														**Ubiquity**
Cereals															
Hordeum vulgare (grain straight)		1	1.5			0.25	1.3	1	0.25					0.3	*
Hordeum vulgare (grain twisted)	0.4	3	1			1	0.7		0.25						*
Hordeum vulgare (grain indet.)	0.4	6	0.5		1	2	3.3		0.25						*
subtotal *Hordeum vulgare* (total grain)	***0.8***	***10***	***3***		***1***	***3.25***	***5.3***	***1***	***0.75***					***0.3***	***64%***
Hordeum vulgare (rachis)	6.4	8	16.5		3	25	44	5.5	20.25					1.3	64%
Triticum dicoccum (grain)	2.0	2	0.5				0.7							0.3	36%
Triticum dicoccum (spikelet fork)	9.2	7	40		3	19	42.7	1.5	39.25					0.6	†
Triticum dicoccum (glume base)	27.2	16	34		1	24	13.3		91.5					1.3	64%
Cereal (grain indet.)	2	1	2		1	2	2.3	2	0.5					1	64%
Cereal culm nodes	0.4					4	1.3								21%
Pulses															
Lens culinaris							0.3							0.7	14%
Large legume	0.4		0.5												14%
Fruits															
Ficus sycomorus ('seed')	72.8	28	30		22	79	402	116	34						57%
Ficus sycomorus (whole fruit frags.)			x					x							14%
Cucumis sp.			2												7%
Ziziphus spina-christi			0.5					0.25							14%
Wild/weed															
Poaceae caryopses (small indet.)	0.4	0.5	2		1	14	0.6	11	2						57%
cf. *Phalaris* sp.	0.8														7%
Lolium cf. *perenne*							0.3	0.5							14%
Poaceae culm node						4			2						14%
Acacia sp. (seed)	0.4						1								
Acacia sp. (seed fragment)	x				x	x	x								29%
Small legumes cf. Trifolieae						4	1.3								14%
Cyperus sp.		1													7%
Fimbrystylis sp.	3.2	9				8	5.3		4						36%
Corchorus sp.			2			1		0.5							21%
Glinus lotoides		1					32		4						21%
Indet.															
		2	4			19	27.3	17.25						3	43%

* Combined under subtotal *Hordeum vulgare* (total grain); † Combined under *Triticum dicoccum* (glume base).

Table 4. Phytolith counts per gram of sediment.

Archaeological sample no.	15	16	17		19	3	23	26	28	42	24	9	274	276	275
Context no.	2022	2025	2026	2084	2092	2016	2104	2128	2130	2182	2113	2044	2098a	2098b	2155
							n per gram sediment								
Single cells															
Smooth LC^{M}	8175	5915	7102	8921	3172	8673	9240	1951	23561	7279			4650	1152	687
Sinuate LC^{M}		539							3534	1456					172
Echinate LC^{M}		3586	6391				3593	1170		5095	6577				69
Dendritic ^{G}	13387	1076	4971	10903		4336	26180	5852	51834	14557	75637		1033		
Dendriform ^{G}										1098					
Trapezoid sinuate/crenate ^{G}			710												
Papillae ^{G}		896	1597	2974			774	967	2356	1456	4385				
Keystone bulliform ^{G}		179		2459	4758		253		1472						102
Bilobate SC^{G}		717		991			1027								
Polylobate SC^{G}															
Cross SC^{G}										728					
Saddle SC^{G}	1487	359	5681			2168				3639					
Saddle topped trapezoid SC^{G}							513								
Rondel SC^{G}	7437	3944	5681	3965		2168	10267	780	8246	8007	8769				137
Indet. SC^{G}	16350		3551				2567								240
Rod smooth LC^{M}	4462		1420	3469			1540		3534	2911					
Assymetrical LC^{M}		359	5681	6938			2053	2341	7068		5481				343
LC with projections on one side ^{M}			7812	6865	6345	6505	1282	1560	4712		4385				618
Stomata ^{M}						1084		184							0
Hair - indet.	3676	5737	22015	17841	15862	21682	13347	5852	41231	9825	18635		14984	6915	1305
Bulliform ^{M}	12631	7889	13493	9912	12690	5420	7187	3511	8246	7279	6206		6717	5762	1511
Cones cf. sedge	4462	3764	2307				4620	1170	1178	728	4385		517	1440	0
Globular decorated indet.			2900			8673		4281		1456			1033		275
Globular echinate cf. palm	14870		2130	1982		6505	5136	3902	2356	9462	3289		3100		206
Globular rugulose cf. palm			710							728					
Dicot elongate ^{D}			353						884	1456				864	278
Tracheids ^{DM}	2975	1434				3202	1790		2356		6577				
Single polyhedral ^{D}	19337	6095	7812	2974			513	4292	10602	5823	25212		2067		1442

Irregular psilate [DM]	15618	3227	4261	7930		8673	1540	390	1178	2911			7750	4610	480
Irregular scrobiculate [D]			1775			4336				990				6339	
Irregular scalloped [D]													517	576	
Tabular scrobiculate [DM]					3172	1084									
Tabular facited edges [DM]			533	505											
Multicells															
Indet. monocot [M]	23799	5464	9232	13876		19875	9240	8974	68326	8189			8220		936
Silica skeleton smooth LC[M]	4282								8540						
Silica skeleton sinuate LC[M]		863	1346				1396								
Silica skeleton echinate LC[M]															
Silica skeleton with rondels SC[G]	2854	144					1625		2062						
Silica skeleton with grass stomata [G]									3534						
Silica skeleton with bilobates SC[G]	1427														
Silica skeleton with saddles SC[G]		215	2051				1396		1178	2736					
Cereal straw/rachis	2975	383					2781	780	10602	337	2035				
Bulliforms [M]	1487	575	12072	2817					2650	6551					202
Silica skeleton assymetrical LC[M]										1368					
Palm multicell		1150	2130	939		28910	3250	3511	1178	2184			470		65
Silica skeleton cf. sedge	1427	288	352	939		1807	1396		2356	674					131
Non ID grass husk silica skeleton (C_3)	5950	1150	673			1807	3251	2340	11780	1368	17294				
Wheat husk silica skeleton		1438	673				2324	780	10602	674	3052				
Barley husk silica skeleton		288					927	1561	5890	2042	5087				
Wild grass husks (C_3)		288				3614	469			674					
Polyhedron multicell (woody dicot leaf)	11900	6902	5681	9912		7227	2324	3902	20027	4779	25433				226
Tracheid multicell [D?]			1420						4712		1525				

[M] = monocot; [G] = grass; [D] = dicot; LC = long cell; SC = grass short cell

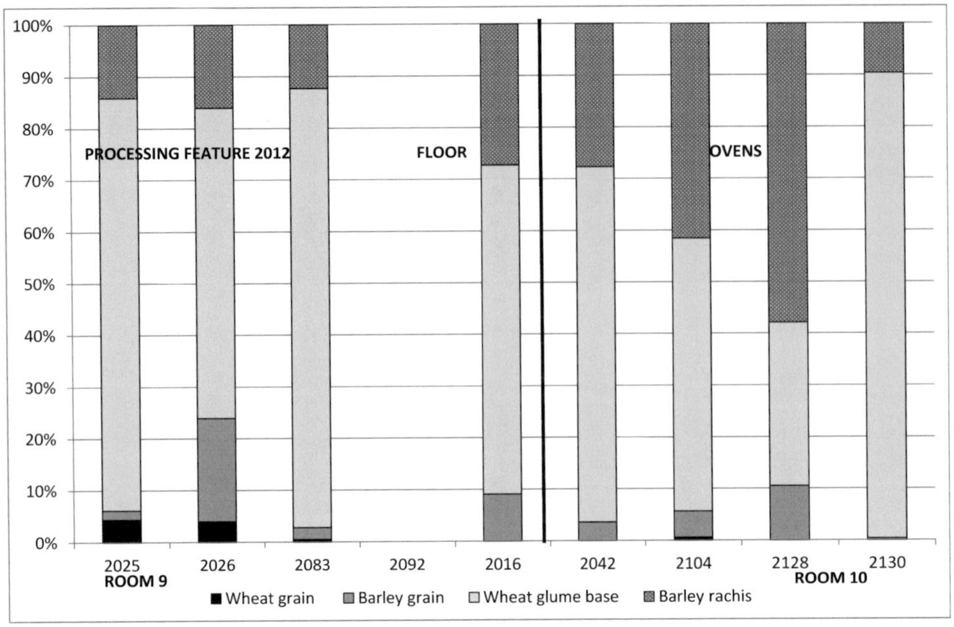

Figure 7. Histogram comparing the charred cereal assemblages from rooms 9 and 10.

categories. However, it is also useful to consider numbers per gram sediment, since percentages sometimes can over-inflate the importance of very small phytolith numbers in samples with low phytolith content.

3.2.1 Single cell phytoliths

A summary of single cell phytoliths, aggregated into monocot, dicot, and palm categories, is shown in Figure 8. These categories are present across all of the samples, but they display some differences in relative proportions and quantities. Sediments from context 2044 (the storage bin in Room 7) had no single cell phytoliths. High relative proportions of dicot phytoliths in comparison to monocots were noted in contexts from rooms 8 and 5, and also in 2092 and in 2016 from Room 9. Palm (globular echinate) phytoliths were intermittently present across the samples. Proportions of different types of grasses are examined in more detail in Figure 9, which compares the relative proportions of different types of grass 'short-cells'. No grass short-cells were present in samples 2092 (Room 9) and 2044 (Room 7) and the two samples from Room 8. Overall, rondel phytoliths (most frequently found in C_3 pooid grasses, which include cereals) were present in the largest proportions and found in 11 contexts. Saddle forms (most frequently found in chloridoid C_4 grasses) were present in more moderate relative proportions and

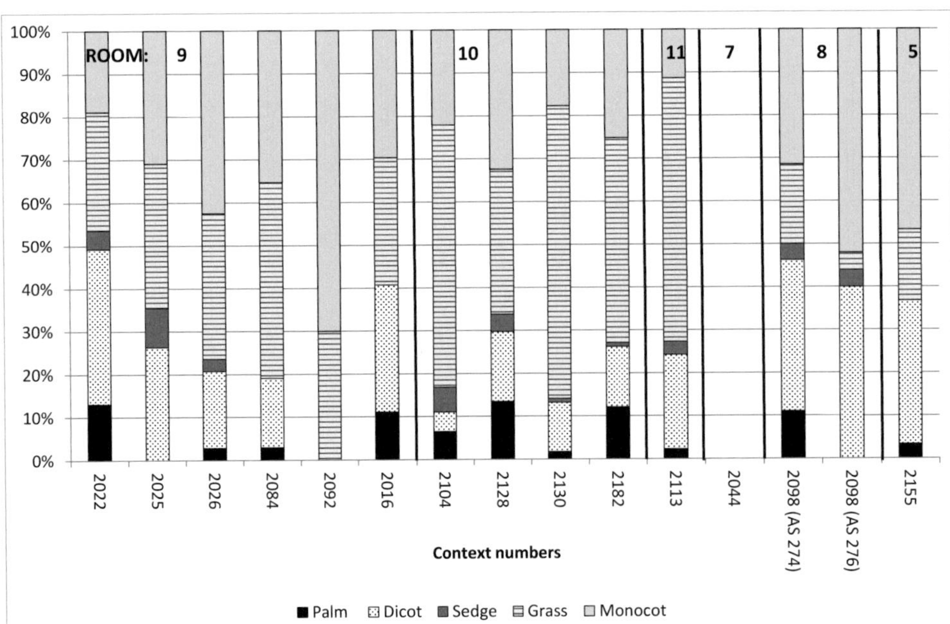

Figure 8. Histogram comparing relative abundances (%) of single cell phytoliths.

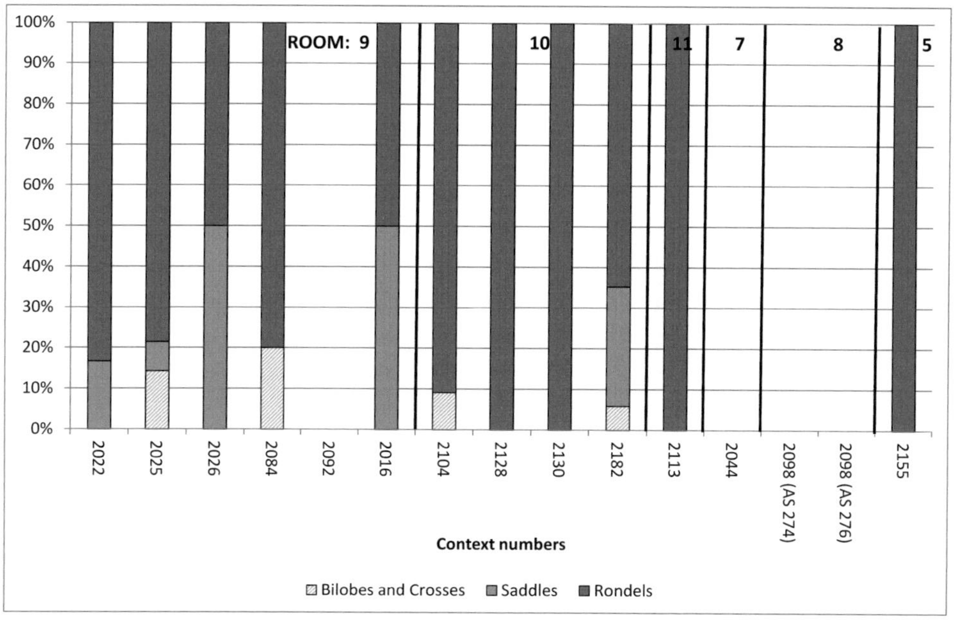

Figure 9. Histogram comparing relative abundances (%) of different grass 'short-cells' phytoliths.

quantities, in fewer samples, but in similar proportions to rondels in samples 2026 and 2016 from Room 9. Bilobate and cross forms (found in panicoid C_4 grasses) were present in the lowest proportions and fewest samples. The histogram in Figure 10 compares numbers of husk single cells (incorporating papillae and dendritic cells [Fig. 6b]) between samples. Grass husks are discussed in more taxonomic detail for the multicell record below; however, the single cell record can provide insight into the presence of grass husks in cases where multicells are not present (for instance, because they are disarticulated). Husk single cells were present in the ashy contexts from rooms 9, 10, and 11 and low/absent in rooms 5, 7, and 8.

3.2.2 Multicell phytoliths

Figure 11a shows a summary of the main categories of multicells present, comparing relative abundances (%) of leaf/stem phytoliths, total husks (cereals + wild + unidentified), and palm and dicot leaf (polyhedrons) (Fig. 12). Sediments from contexts 2092 (Room 9), 2044 (Room 7), and 2098 (AS 276) (Room 8) contained no multicell phytoliths. In the other samples, leaf and stem phytoliths

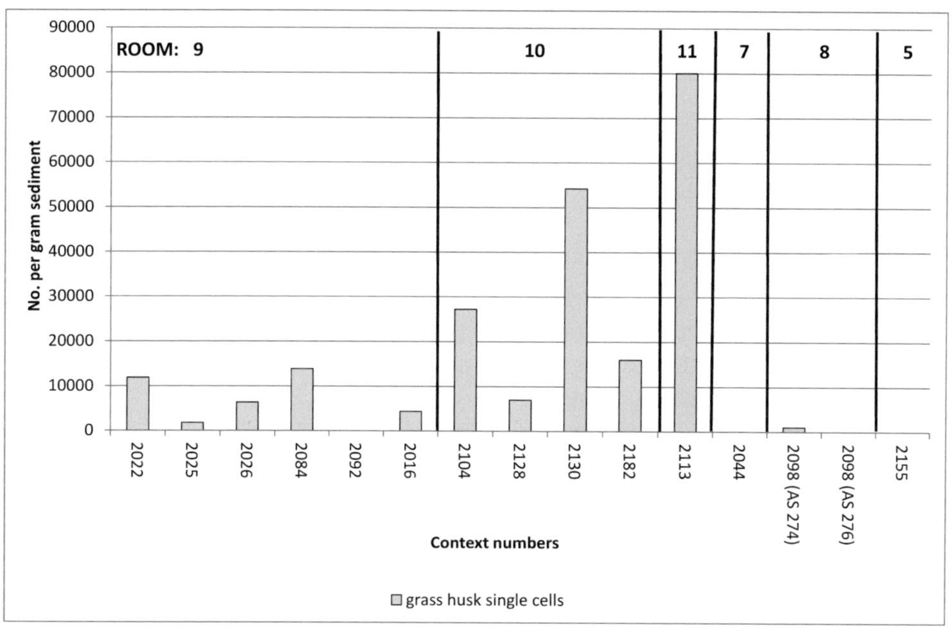

Figure 10. Histogram comparing numbers per gram sediment of grass husk single cell phytoliths.

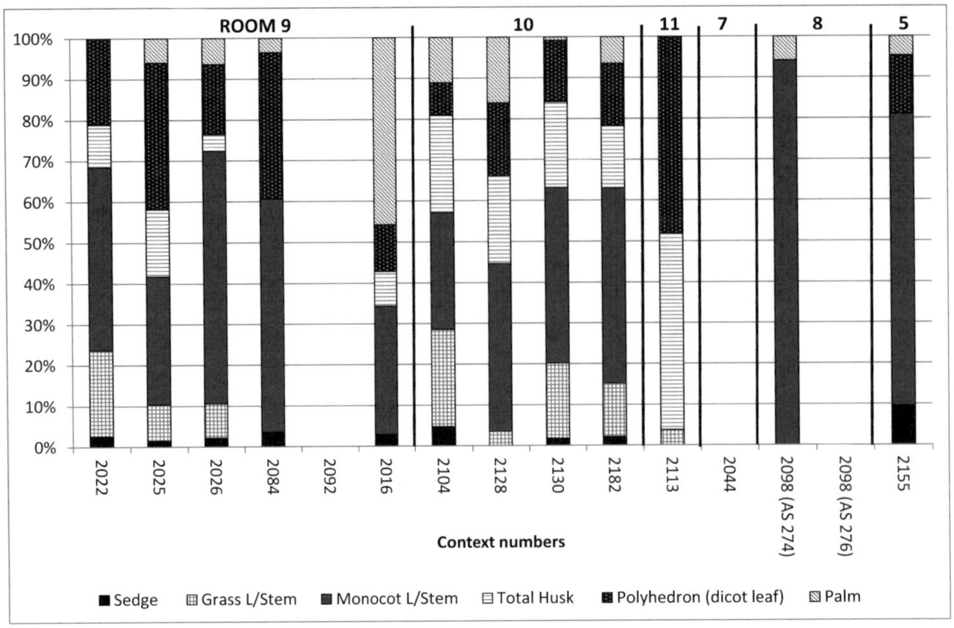

Figure 11a. Histogram comparing relative abundances (%) of multicell phytoliths.

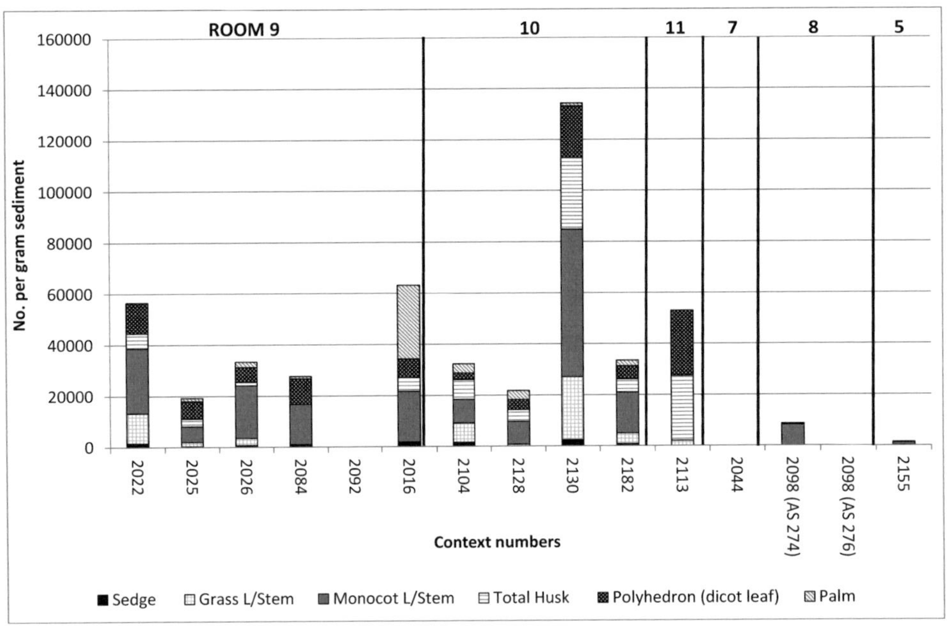

Figure 11b. Histogram comparing numbers per gram sediment of multicell phytoliths.

were ubiquitous. Husks were variably present, with the largest relative proportions and occurrence in samples from the oven room and the possibly related ashy deposit in adjacent Room 11, whilst rooms 5, 7, and 8 had no husk multicells present. Proportions of palm and dicot leaf phytoliths were variable among contexts, with ashy floor deposit 2016 (Room 9) having the highest proportion of palm multicell phytoliths. Several samples had moderate to high proportions of dicot leaf phytoliths, especially context 2113 (Room 11). Figure 11b shows the numbers of multicell phytoliths per gram of sediment, and thus more clearly shows the small numbers of phytoliths in some samples, for instance, in that from ashy context 2155. Grass husks in the multicell record were categorized as wheat (Fig. 13), barley, non-identifiable (C_3), wild (C_3), or panicoid. The relative proportions (%) among these categories are shown in Figure 14.

No grass husk multicell phytoliths were present in the samples from rooms 5, 7, and 8. Within Room 9, proportions of husk multicells varied; two samples had high proportions from wheat, one contained small proportions from barley, two had no husk multicells, and one con-

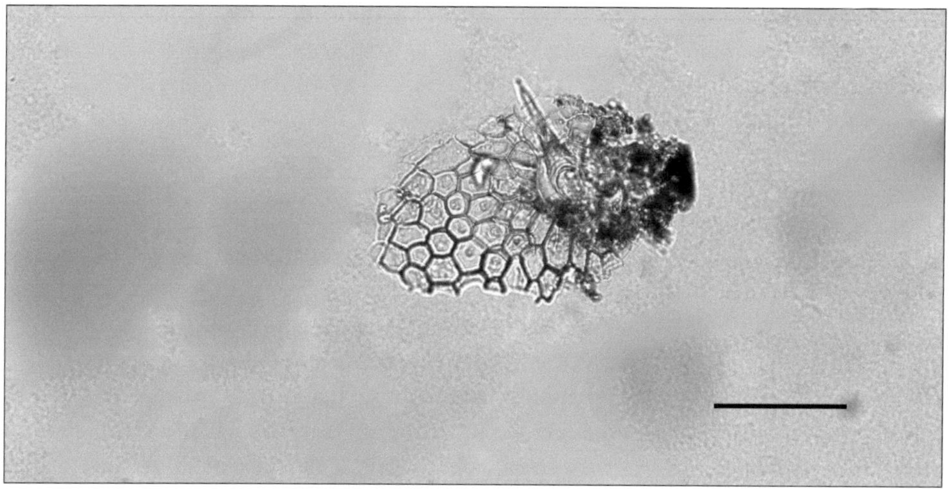

Figure 12. Multicell phytolith cf. dicot leaf from oven context 2130, scale bar 100 microns. (Image: P. Ryan; © Trustees of The British Museum.)

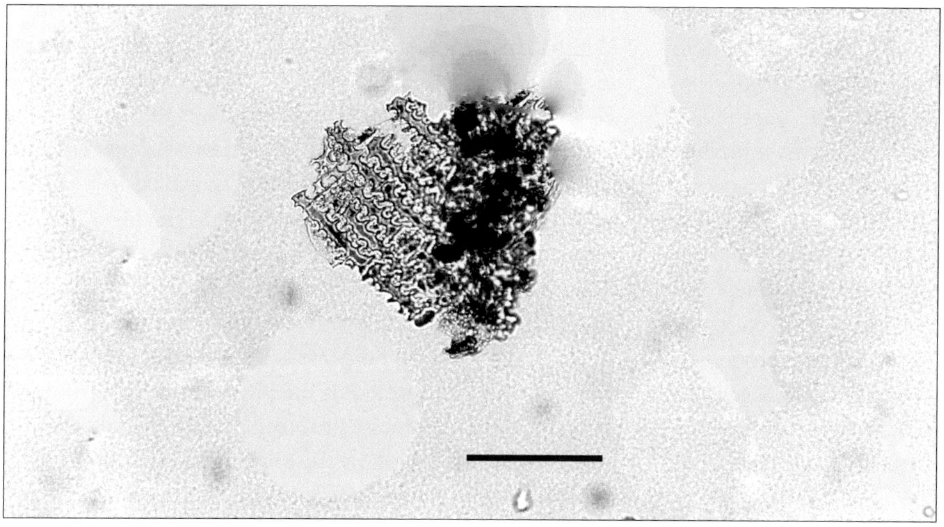

Figure 13. Multicell phytolith cf. wheat husk (*Triticum* sp.) from oven context 2130, scale bar 100 microns. (Image: P. Ryan; © Trustees of The British Museum.)

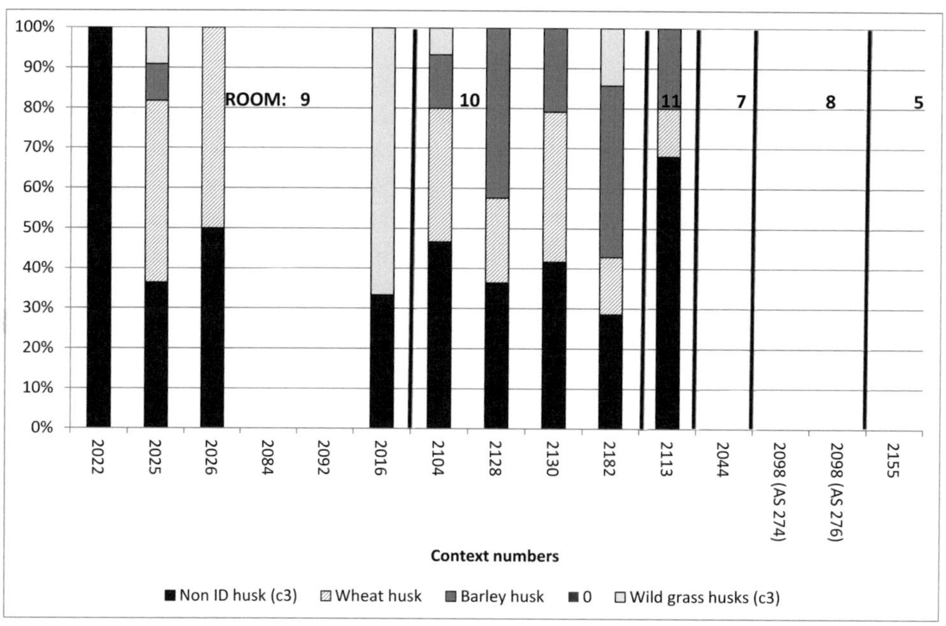

Figure 14. Histogram comparing relative abundances (%) of grass husk multicell phytoliths.

tained unidentifiable husk forms. In contrast, wheat and barley husk phytoliths were present in all of the samples from Room 10, and the ashy sample from Room 11 had a similar composition to that of Room 10. Two samples from Room 10 and one from Room 9 contained small proportions of wild grass husk phytoliths, and one sample from Room 9 was notable for having the highest proportion of wild grass husks (which were from C_3 grasses), in combination with zero cereal phytoliths.

3.2.3 Phytolith results by room

Room 9 (grain processing room): Several differences were observed among the different samples analysed from Room 9, and between these samples and those from the other villa rooms. In comparison with all the other samples analysed, samples from 2055 and 2056 (different areas of emplacement for grinding stones 2012) had the highest relative proportions of wheat husk phytoliths, whilst ashy floor sample 2016 had the highest relative proportions of phytolith multicells from wild grass husks and also from palms. Floor context 2092 forms a further contrast within Room 9, being the only sample to have no phytoliths from grass husks and no multicell phytoliths.

Room 10 (oven room): All of these samples were from ashy deposits within or near the cylindrical ovens. The overall phytolith compositions among samples were similar, with only slight variations. All contained phytoliths from monocot leaves/stems, dicots, palms, and cereals. Two also contained phytoliths from wild grasses (which included pooid grass husk multicells and single cells from C_4 grasses of uncertain plant part).

Room 11 (small room adjacent to the oven room): The grass husk composition of ashy context 2113 was similar to deposits analysed from Room 10. However, there were differences among proportions of other phytoliths present. Only small proportions of monocot leaves/stems were present, in contrast with comparatively high proportions of dicot leaf phytoliths.

Room 7 (courtyard): The sample analysed from a storage bin context in courtyard Room 7 did not contain any phytoliths.

Rooms 5 and 8: In the single cell record, these samples contained mainly dicot phytoliths and morphologies identifiable only as monocot. Fewer grass single cell phytoliths were identified than in the other samples, and grass husk single cells occurred in only one sample (2098), and in negligible numbers. Small numbers of multicells were present in two of the samples (ashy floor sample 2098 and hearth 2155), and these were dominated mainly by morphologies identifiable as monocot leaf/stem, with smaller proportions from palms, and in the hearth sample, also from sedges and woody dicot leaves.

4. Discussion

4.1 Comparison of macroremains and phytoliths

The macroremains have enabled the identification of a wider range of plant taxa, whilst the phytoliths have provided information about certain plant parts less frequently identified in the macrobotanical record, such as leaves and stems. Diagnostic husk phytoliths are found in the lemma, palea, and glume (no phytoliths are present within grass seeds). Both datasets have provided information about cereals and wild grasses. Sometimes the presence of barley chaff does not correlate as clearly as that of wheat between the datasets because barley rachises (the stems within the cereal ear) do not produce husk phytoliths, but, rather, phytoliths associated with cereal stems. Wild C_3 grasses were identified in both the macrobotanical and phytolith record. However, generally, small wild grass caryopses were often very highly charred in the macrobotanical record from the villa; here the phytolith record provided additional information about the types of wild grasses present. Small quantities of C_4 grasses were noted in the phytolith record from the villa, particularly from chloridoid grasses, but also (more occasionally) from panicoid grasses. This is interesting, as chloridoid grasses are generally more arid-adapted (Twiss 1992). Occasional charred chloridoid grass seeds have also been identified in deposits analysed from other buildings at Amara West. Phytolith information also provided plant evidence for contexts where few or no charred macroremains had survived (notably the ashy deposit in Room 8) and confirmed in other instances the absence of plant remains (such as in the storage context).

Processing Room 9 and oven Room 10 were notable for the presence of abundant cereal remains, particularly chaff, possibly reflecting the activities of crop processing and the burning of crop processing waste. Some possible differences between the grain processing structure contexts (Room 9) and the oven room (Room 10) were noted. Both sets of deposits contained, overall, similar quantities of charred wheat chaff per litre sediment. However, both the phytolith and charred record showed greater relative proportions of chaff from wheat than from barley within deposits from the grain processing structure contexts (Room 9) in comparison with deposits from the oven room (Room 10). In addition, wheat grains were recovered with greater frequency from the grain processing structure than from the ovens, whilst one of these grain processing samples also had the highest numbers and relative proportions of barley grains. One of the two Room 9 floor samples had low/absent plant content in both the macro- and microbotanical record. The other sample, which was ashy, contained few macroremains, but the phytolith data indicated comparatively high relative proportions of both wild grass husks and palms. The burnt wild grasses from the floor in the processing room possibly represent a distinctive episode and location of burning crop-cleaning debris. The high relative proportion of multicell phytoliths from palms also marked this floor deposit as distinct from the other ashy deposits. Charred cereal remains around the processing feature may also relate to parching; however, experimental work in Egypt suggested that the process of parching prior to de-husking was not required due to aridity (Nesbitt & Samuel 1996). Another possibility is that ash was spread around quern settings deliberately, although the potential purpose of this is unclear. Miller (1987) has argued that, because ash has insecticidal properties, ash found beneath saddle querns at Tell el-Amarna may have been deliberately spread for this purpose.

The presence of wheat chaff in processing Room 9 and oven Room 10 suggests that emmer was stored in spikelet form. During threshing of hulled cereals the ears are broken down into individual spikelets. They can be stored in this form, but they then require a second processing stage to release the grains from the husk. For emmer wheat, the by-product of this second processing stage comprises light, papery lemmas and palea parts and more robust spikelet forks (or individual glume bases), and it is these parts that survive in the charred record. In contrast, during threshing, barley breaks into spikelets consisting of grains within a tightly fused lemma and palea, and segments of barley rachis. Thus, the presence of barley rachis segments is more ambiguous than that of wheat husk chaff (glume bases/spikelet forks). Barley rachis chaff may have been deliberately collected off-site to serve as tinder for fire-lighting. Another possibility is that barley ears were harvested and stored in ear form. Barley may also have been associated with dung. Much of the barley was well preserved and would not have survived well after animal digestion, so perhaps cereal remnants were incorporated into dung deposits in animal pen floors through trampling. Dung pellets have been identified in some charred deposits on the site, but pellets have not been recovered (so far) from the villa samples—though this may reflect burning temperatures rather than real absence. Barley remains may also derive from their processing for food; some grains displayed hull (lemma and palea) removal, and barley grains were recovered from all the grinding emplacement samples. It is interesting that there were slight differences in proportions of barley to wheat chaff between rooms 9 and 10, suggesting some possible different routes into these contexts for wheat and barley.

In both rooms 9 and 10, phytoliths from monocot leaves/stems (including grasses and sedges), woody leaves, and palms most probably derived from the ashy remains of fuel. Some of the small wild seeds (including the grasses) may have entered the record with leaves/stems in connection with fuel or may have been present as crop weeds. An ashy context from Room 11 (near oven Room 10) was analysed for comparison with the oven deposits. Although no seeds were preserved in the ash, it did contain cereal phytoliths and, in contrast to the oven deposits, high proportions of phytoliths from woody leaves and a few from monocot leaves/stems. All this suggests the fuel was of a different composition and that these ashes were not necessarily spread from the adjacent room.

The sample from the storage bin sediments (Room 7) was sterile in both datasets, indicating the deposit was purely wind-blown sand and suggesting that the bin had been emptied prior to abandonment. Such a practice would fit within the broader picture of careful abandonment suggested by the comparatively few artefactual finds and clean floors of the villa. The samples analysed from rooms 5 and 8 contrasted with those analysed from other rooms. Cereals were absent in the record from Room 8, and Room 5 contained negligible numbers of charred cereals and no evidence of cereals in the phytolith record. Phytoliths present in these two rooms probably derived principally from burnt fuel. The southern part of the building seems to be more associated with cereal processing, with ashy samples from the north area of the villa predominantly reflecting only fuel remains. The hearth in Room 5 would have been one of the focal points to life in the house, providing heat for cooking and warmth. The archaeobotanical record thus confirms interpretations based on the architectural layout and presence of features.

4.2 Comparison with other site contexts

The villa analyses form part of a pilot study and of ongoing research into the archaeobotanical evidence from smaller houses within the north-western part of the town (E13.3-N and E13.3-S); a circular building (E12.11); and other samples, such as from roofing fragments (Ryan et al. 2012; Ryan & Spencer 2013). There are no burnt buildings on-site, and charred plant remains most probably represent fuel and other plant materials connected to cooking or debris. Exceptions include desiccated fragments of textile and basketry and wood artefacts preserved in cemetery contexts. Preservation via desiccation in later time periods (in cemetery contexts) may relate to increased aridity. Overall findings were similar to samples analysed from other site contexts, but there are also some initial distinctions. So far, there is also a greater proportion of barley to wheat in the smaller houses, E13.3-N and E13.3-S, particularly in the chaff record (45% wheat, 55% barley), whilst another notable difference was the presence of flax seeds in E13.3-S (Ryan et al. 2012). Some other taxa present elsewhere in the charred macrobotanical record but not yet identified in the villa include *Hyphaene thebaica* Mart. (doum palm), *Citrullus lanatus* (Thunb.) Matsum & Nakai (watermelon), and the closely related wild relative *Citrullus colocynthis* (L.) Schrad. (colocynth).

A key difference between the villa and the smaller houses seems to be that storage bins were not present in the latter. It is possible that the smaller houses received cereal supplies either from the larger houses or from storage magazines under the centralized control of the town administration and then maybe kept them in perishable storage containers. The different proportion of wheat to barley grains and chaff between the villa and smaller houses may also suggest variable household access to cereal resources. However, some differences in charred cereal assemblages may be context-related, since the charred deposits analysed from the smaller houses are so far all from ovens and hearths. Differences may also reflect varied taphonomic pathways for grains and chaff in relation to, for example, crop processing debris collection and disposal practices or fuel use. Some of the smaller houses in the town that are contemporary with the villa (E13.4, E13.5, E13.6) were provided with suites of cylindrical ovens and grinding emplacements, though on a smaller scale than at villa E12.10. All houses were provided with a circular hearth, but other approaches to foodways are evident within the town. Houses E13.3-N and E13.3-S seemed to have shared a space that housed ovens and grinding emplacements. This space (E13.13) was accessible from the front door of each house; such an organization has implications in terms of privacy and social interactions within the ancient town. Villa D12.5 (currently being excavated), to the south of and of similar scale to villa E12.10, was also provided with a food processing suite, though it was arranged differently: a long room with three circular silos for possible grain storage, followed by rooms dedicated to grinding emplacements and ovens. One silo is provided with a mud plaster floor; the other is paved with schist and coated in white plaster, perhaps suggesting differential use patterns.

4.3 Diet and plant use at villa E12.10 in the context of New Kingdom Egypt

The evidence thus far suggests that the cereals were Pharaonic winter crops. Bread and beer were staple Egyptian foods and were also common ritual offerings (Samuel 2001). There is not yet any evidence for local African grass exploitation, and this is clear in both the macrobotanical and phytolith record. However, analyses from new buildings or site levels at Amara West may alter this picture. Low numbers of pulses found at Amara West fit with similar observations elsewhere in New Kingdom Egypt (Murray 2000; Stevens & Clapham 2010).

Circular structures interpreted as ovens and features interpreted as grinding emplacements at Amara West are both similar to those excavated in Egypt, including at Tell el-Amarna (Samuel 2000). Like at Amara West, there is also evidence for storage facilities within various categories of elite homes at Tell el-Amarna (Kemp 1994). In contrast, the villas at Amara West were not situated within a walled compound, which, at Tell el-Amarna, included external subsistence activity areas such as circular grain silos, kitchen areas, and animal pens (Kemp 2006: 327-329). There are no stone mortars (for cereal de-husking) at Amara West such as those found at the workmen's village at Amarna or Deir el-Medina in Egypt (Samuel 2000); it is possible that wooden implements were used instead.

5. Conclusions

The different lines of archaeobotanical data are useful for cross-checking results and providing different categories of information about plant use. Archaeobotanical remains from Amara West are contributing to a better understanding of the late second millennium to early first millennium BC in Nubia (Ryan et al. 2012). The identification of wood, charcoal, and textile fibres from the cemeteries and settlement is ongoing, and the phytolith and seed record will be usefully combined with the charcoal data for a fuller view of fuel exploitation and differences in fuel use among individual ovens and hearths. Future aims also include further comparison between the villa and other site areas, with excavations ongoing both inside the walled town and in the extramural suburb. Temporal analyses will be important for investigating the impact of increasing local aridity on agriculture and plant use strategies.

Acknowledgments

Fieldwork at Amara West would not be possible without the permission and support of the National Corporation of Antiquities and Museums (Sudan), with particular thanks to Abdel Rahman Ali Mohamed Hassan Hussein Idris, Salah Mohamed Ahmed, Shadia Abdu Rabo, and Mohamed Saad. Excavations in villa E12.10 were supervised by Neal Spencer, An Van Camp, and René Kertesz, whilst the entire ceramic assemblage was studied by Marie Millet. Thanks also to Marie Vandenbeusch, Mary Shepperson, and Matthew Dalton from the Amara West excavation team. Thanks also to Dorian Fuller for helpful advice and to Wakehurst Place, the Royal Botanic Gardens Kew, for providing reference material. The Amara West research project is made possible through the generous funding of The Leverhulme Trust (research project *Health and diet in occupied Nubia through political and climate change*), and also through the British Academy and the Fondation Michela Schiff Giorgini. For further information on the project, visit www.britishmuseum.org/AmaraWest.

References

Andrews F.W., 1956. *The Flowering Plants of the Anglo-Egyptian Sudan. Vol. III. Compositae-Gramineae.* Bungel and Co., Arbroath.

Binder M., 2011. The 10th-9th century BC—new evidence from Cemetery C of Amara West. *Sudan & Nubia* 15, pp. 39-53.

Binder M., Spencer N. & Millet M., 2011. Cemetery D at Amara West: The Ramesside Period and its aftermath. *British Museum Studies in Ancient Egypt and Sudan* 16, pp. 47-99.

Boulos L., 1999. *Flora of Egypt. Vol. 1. Azollaceae-Oxalidaceae.* Al Hadara Publishing, Cairo.

Boulos L., 2000. *Flora of Egypt. Vol. 2. Geraniaceae-Boraginaceae.* Al Hadara Publishing, Cairo.

Boulos L., 2002. *Flora of Egypt. Vol. 3. Verbenaceae-Compositae.* Al Hadara Publishing, Cairo.

Boulos L., 2005. *Flora of Egypt. Vol. 4. Monocotyledons (Alismataceae-Orchidaceae).* Al Hadara Publishing, Cairo.

Braun M. & Burgstaller H., 1991. *Common Weeds of Central Sudan.* Margraf, Weikersheim.

Bullock P., Fedroff N., Jungerius A., Stoops G. & Tursina T., 1985. *Handbook for Soil Thin Section Description.* Waine Research,Wolverhampton.

Edwards D.N., 2004. *The Nubian Past: An Archaeology of the Sudan.* Routledge, London.

Fuller D.Q., 2004. Early Kushite agriculture: Archaeobotanical evidence from Kawa. *Sudan & Nubia* 8, pp. 70-74.

Hillman G., Madeyska E. & Hather J., 1989. Wild plant foods and diet at Late Paleolithic Wadi Kubbaniya. In: Wendorf F., Schild R. & Close A.E. (eds.), *The Prehistory of Wadi Kubbaniya. Vol. 2. Stratigraphy, Paleoeconomy, and Environment.* Southern Methodist University Press, Dallas, pp. 162-242.

Kemp B.J., 1994. Food for an Egyptian city. In: Luff R. and Rowley-Conwy P. (eds.), *Whither Environmental Archaeology?* Oxbow Monograph 38. Oxbow, Oxford, pp. 133-153.

Kemp B.J., 2006. *Ancient Egypt: Anatomy of a Civilization.* Routledge, London.

Miller R., 1987. Appendix: Ash as an insecticide. In: Kemp B.J. (ed.), *Amarna Reports IV*. Egypt Exploration Society, London, pp. 14-16.

Murray M.A., 2000. Cereal production and processing. In: Nicholson P.T. & Shaw I. (eds.), *Ancient Egyptian Materials and Technology*. Cambridge University Press, Cambridge, pp. 505-536.

Nesbitt M. & Samuel D., 1996. From staple crop to extinction? The archaeology and history of the hulled wheat. In: Padulosi S., Hammer K. & Heller J. (eds.), *Hulled Wheat.* [Proceedings of the 1st International Workshop on Hulled Wheats.] Promoting the Conservation and Use of Underutilized and Neglected Crops 4. International Plant Genetic Resources Institute, Rome, pp. 40-99.

Piperno D.R., 2006. *Phytoliths: A comprehensive Guide for Archaeologists and Paleoecologists.* Alta Mira Press, Lanham.

Rosen A.M., 2005. Phytolith indicators of plant and land use at Çatalhöyük. In: Hodder I. (ed.), *Inhabiting Çatalhöyük: Reports from the 1995-99 Seasons. Vol. 4.* McDonald Institute for Archaeological Research, Cambridge, pp. 203-212.

Ryan P., Cartwright C.R. & Spencer N., 2012. Archaeobotanical research in a pharaonic town in ancient Nubia. *British Museum Technical Research Bulletin* 6, pp. 97-106.

Ryan P. & Spencer N., 2013. Diet and plant-use at Amara West. *Egyptian Archaeology* 42, pp. 18-20.

Ryan P., Weisskopf A. & Rosen S.A., 2011. Sediments and microartifacts from the Camel site. In: Rosen S.A. (ed.), *An Investigation into Early Desert Pastoralism: Excavations at the Camel Site, Negev.* Cotsen Institute of Archaeology Press at UCLA, Los Angeles, pp. 155-166.

Samuel D., 2000. Brewing and baking. In: Nicholson P.T. & Shaw I. (eds.), *Ancient Egyptian Materials and Technology.* Cambridge University Press, Cambridge, pp. 537-577.

Samuel D., 2001. Bread. In: Redford D.B. (ed.) *The Oxford Encyclopedia of Ancient Egypt.* Oxford University Press, Oxford, pp. 196-198.

Smith S.T., 2003. *Wretched Kush: Ethnic Identities and Boundaries in Egypt's Nubian Empire*. Routledge, New York.

Spataro M., Millet M. & Spencer N., 2014. The New Kingdom settlement of Amara West (Nubia, Sudan): Mineralogical and chemical investigation of the ceramics. *Archaeological and Anthropological Sciences*, 1-23. doi 10.1007/s12520-014-0199-y

Spence K., 2004. The three-dimensional form of the Tell el-Amarna house. *Journal of Egyptian Archaeology* 90, 123-152.

Spencer N., 2009. Cemeteries and a Ramesside suburb at Amara West. *Sudan & Nubia* 13, pp. 47-61.

Spencer N., 2010. Nubian architecture in an Egyptian town? Building E12.11 at Amara West. *Sudan & Nubia* 14, pp. 15-24.

Spencer N., 2014. Amara West: Considerations on urban life in occupied Kush. In: Anderson J.R. & Welsby D.A. (eds.), *The Fourth Cataract and Beyond*. [Proceedings of the 12th International Conference for Nubian Studies.] Leuven, pp. 457-485.

Spencer N., 2015. Amara West: House and neighbourhood in Egyptian Nubia. In: Müller M. (ed.), *Household Studies in Complex Societies: (Micro)archaeological and Textual Approaches*. Oriental Institute Seminars 10. Oriental Institute, University of Chicago, Chicago, pp. 169-210.

Spencer N. & Hay S., 2012. Amara West: Remote sensing at a pharaonic town in northern Sudan. In: Johnson P. & Millett M. (eds.), *Archaeological Survey and the City*. University of Cambridge Museum of Classical Archaeology Monographs 2. Cambridge University Press, Cambridge, pp. 176-201.

Spencer N., Macklin M. & Woodward J., 2012. Reassessing the abandonment of Amara West: The impact of a changing Nile? *Sudan & Nubia* 16, pp. 37-43.

Spencer P., 1997. *Amara West. Vol. I. The Architectural Report*. Egypt Exploration Society, London.

Spencer P., 2002. *Amara West. Vol. II. The Cemetery and the Pottery Corpus*. Egypt Exploration Society, London.

Stevens C.J. & Clapham A., 2010. The botanical samples. In: Kemp B. & Stevens A. (eds.), *Busy Lives at Amarna: Excavations in the Main City (Grid 12 and the House of Ranefer, N49.18). Vol. 1. The Excavations, Architecture and Environmental Remains*. Egypt Exploration Society Excavation Memoir 90. Egypt Exploration Society and Amarna Trust, London, pp. 427-443.

Täckholm V., 1974. *Students' Flora of Egypt*. Cairo University, Beirut.

Terry R.D. & Chilinger G.V., 1955. Comparison charts for visual estimation of percentage composition. *Journal of Sediment Petrology* 25, pp. 229-234.

Tietze C., 1985. Amarna: Analyse der Wohnhäuser und soziale Struktur der Stadtbewohner. *Zeitschrift für ägyptische Sprache und Altertumskunde* 112, pp. 48-84.

Twiss P. C., 1992. Predicted world distribution of C_3 and C_4 grass phytoliths. In Rapp Jr. G. & Mulholland S.C. (eds.), *Phytolith Systematics: Emerging Issues*. Plenum Press, New York, pp. 113-128.

Pearl millet, *Pennisetum glaucum* (L.) R.Br. ssp. *glaucum*, in the Dakhleh Oasis, Egypt

Ursula Thanheiser

VIAS-Archaeobotany
Vienna University
Althanstrasse 14
1090 Vienna, Austria

E-mail: ursula.thanheiser@univie.ac.at

Stefanie Kahlheber

Buderusstrasse 1
65549 Limburg, Germany

E-mail: kahlheber@gmail.com

Tosha Dupras

Department of Anthropology
University of Central Florida
Orlando, FL 32816 USA

E-mail: tosha.dupras@ucf.edu

Abstract. Although pearl millet is an important grain crop, little is known about its domestication and subsequent dispersal. The numerous archaeobotanical remains from Kellis, a Roman settlement in the Dakhleh oasis in the Western Desert of Egypt, serve as a trigger to explore its history as well as its role in the food chain in oasis agriculture.

Keywords: pearl millet, origin, dispersal, nutrition, Egypt

1. Introduction

Pearl millet—*Pennisetum glaucum* (L.) R.Br. ssp. *glaucum*—is an important grain crop. Together with wheat, barley, maize, rice, and sorghum, it feeds the world's growing population. However, little is known of the time of arrival and subsequent cultivation of this African crop in northeastern Africa and the nutritional role it eventually gained in an area where Near Eastern and Mediterranean crops have fed the population for several millennia.

Pearl millet is one of the most frequent and abundant archaeobotanical finds at the Roman settlement of Kellis, in the Dakhleh oasis in the Western Desert of Egypt. It is present in all domestic areas throughout the village, from the beginning of its occupation in the second century AD to its abandonment in the late fourth century, and seems to have played an important role in the food chain. But unlike other provisions, such as wheat and barley, poultry, olive oil, and new wine, it is not mentioned in the Kellis Agricultural Account Book, a record of dues compiled by the overseer of an estate in the late fourth century AD (Bagnall 1997: 35-46) and therefore appears to have been grown for the tenants' personal utilisation and not as a cash crop.

2. Pearl millet biology and cultivation

As an African cultigen of Sahelian origin, pearl millet is well adapted to dry and warm climates; it therefore has its main distribution today in the arid and semi-arid zones of tropical Africa and the Indian subcontinent. In these areas pearl millet grains are traditionally used for food and fodder, and its stalks for fodder, fuel, and fencing. Areas where pearl millet is grown can be characterised by high temperatures, low and erratic precipitation, poor soil fertility, and high disease and pest pressure. Due to its high adaptability and stress resistance, pearl millet can grow in marginal areas where other crops are not successful. It therefore serves as a base-line food security (National Research Council 1996: 77-91).

Pearl millet is an erect annual grass with good tillering habit. The number of tillers produced by a single plant depends on the genotype and on environmental conditions. Its main stem can attain a height of 2–4 m, sometimes up to 6 m. Pearl millet has an extensive root system, penetrating deeply into the ground. Above the soil surface the plant develops so-called crown roots, which support the plant and prevent it from lodging. The flowers are arranged in a panicle with an average length of 25 cm. There are varieties where panicles attain a length of 200 cm. The panicle consists of a central axis (rachis), which is covered with soft hairs. Fascicles on rachillae are arranged in spirals on the rachis. The density of the fascicles and the length of rachillae determine the compactness of the panicle. Each fascicle contains spikelets surrounded by a whorl of bristles (involucrum). The grain is a caryopsis with highly variable shape, ranging from globular to nearly conical. Its size depends on its position on the panicle, with the largest grains at the base and the smallest ones at the apex. The weight of 1000 grains ranges from 3 to 15 g (Andrews & Kumar 2006; Mangat et al. 1999; Oyen & Andrews 1996; Stoskopf 1985: 434-440).

The properties of the edible portion of the grain depend on the cultivar and, especially for the trace elements, on the environment. Pearl millet is an important source of carbohydrates, but it also provides fat and all the essential amino acids, and is a richer source of protein, calcium, phosphorus, and iron than some of the other cereals of world importance (Andrews & Kumar 2006; Balole & Legwaila 2006; Ceccarelli & Grando 2006; Singh & Nainawatee 1999).

Characteristic of pearl millet is an extensive and fast-growing root system, which results in high drought tolerance. Although it can be grown in a variety of soils (except under waterlogged conditions), it prefers light loamy and sandy soils. Soil salinity adversely affects germination, and soil crust formation due to rainfall immediately after sowing poses a serious impediment to seedling emergence. Germination is rapid at high temperatures. High temperatures are also required for rapid growth, with an optimal temperature of 33–34°C. Water requirements are low. The northward extension of pearl millet cultivation into the drier parts of sub-Saharan Africa is usually defined by the 250 mm isohyet, but under otherwise favourable conditions, pearl millet can also grow in areas with an annual precipitation as low as 125 mm. However, it responds well to irrigation. Water stress has its most adverse effects during early plant establishment and during the flowering and grain-filling period. The growth duration of pearl millet is variable, ranging from 50 to 280 days, and is dependent on the cultivar and on environmental conditions. Two cultivar groups can be distinguished within ssp. *glaucum*, namely, *gero* and *maiwa*. The first has a growth duration of about 90 days and is therefore adapted to a short rainy season. It exhibits little photosensitivity. *Maiwa*-cultivars on the other hand grow slowly, and flowering is induced by short day length (< 12 h); they are mainly cultivated in regions with a longer rainy season. In the oases of the Sahara desert cultivars that are photoperiod insensitive are grown under irrigation (Andrews & Kumar 2006; Burkill 1994: 316; FAO; Mangat et al. 1999). Like maize, sorghum, and other tropical crops, pearl millet is a C_4 plant; it thus has a photosynthetic adaptation to short days, enabling the plant to metabolize CO_2 more efficiently.

Pearl millet is usually sown directly into the ground. After germination, an extensive root system is established and the growth rate of the parts above ground is comparatively slow in the beginning. Generous spacing between single plants is necessary in marginal areas with little rainfall. This spacing is accomplished by thinning the crop at first weeding. This way plant density can be adjusted to average rainfall or the availability of water for irrigation, as well as to soil fertility. The time to flower initiation is the main factor determining the length of the life cycle of a cultivar (Andrews & Kumar 2006; National Research Council 1996: 93-109).

Pearl millet is a dual crop, producing grain as well as greenery for livestock feeding and fencing. Traditional varieties tend to be tall, leafy plants that perform best when spaced apart and that produce massive amounts of greenery. The harvest index therefore is often below 20 per cent, which means that more than 80 per cent of the plant above ground is stalk and leaf. Crop productivity of land-races lies between 500 and 600 kg/ha grain yield, in marginal areas often only 250 kg/ha, and is rather poor compared with other grain crops. The low yield, however, is compensated for by the fact that pearl millet will grow and produce grain even under the most adverse conditions (National Research Council 1996: 93-109; Stoskopf 1985: 433).

3. Origin, domestication, and distribution of pearl millet

Pearl millet is the essential staple crop of semi-arid sub-Saharan Africa, particularly so north of the equator. The cultivated form derives from *Pennisetum glaucum* ssp. *violaceum* (Lam.) van der Zon, which occurs in the Sahel and the southern Sahara (Brunken et al. 1977). Morphological changes involved in the domestication syndrome affect mainly the inflorescence. Most significant is the loss of a callus at the base of the involucrum, resulting in a non-shattering of diaspores (plant dispersal units, i.e. in this case the grain kernels covered by the involucra). For a distinction of wild and domesticated forms in the archaeobotanical material, the differences in caryopsis shapes, dimensions, and the length of involucrum stalks offer useful diagnostic features.

Although it was an important crop in West Africa and India for millennia, our knowledge on the location, timing, and cultural circumstances of pearl millet domestication is far from complete; in fact, it is to a large degree still hypothetical. Based on the modern distribution of wild pearl millet, Brunken (1977), Harlan (1971), and others have placed the domestication area at the southern fringe of the Sahara, in a belt across the continent. More recently, enzyme studies lead Tostain (1998) to suggest an area in southern Mauritania, Senegal, and eastern Mali as a centre of early domestication (Fig. 1) from where the crop could have spread across the continent.

Archaeobotanical dates for pearl millet suggest a West African origin as well (see compilations in Kahlheber & Neumann 2007; Manning et al. 2011). The oldest African finds so far originate from the Tilemsi valley and date to the third millennium BC (Manning et al. 2011). In the early second millennium BC, pearl millet is extant in

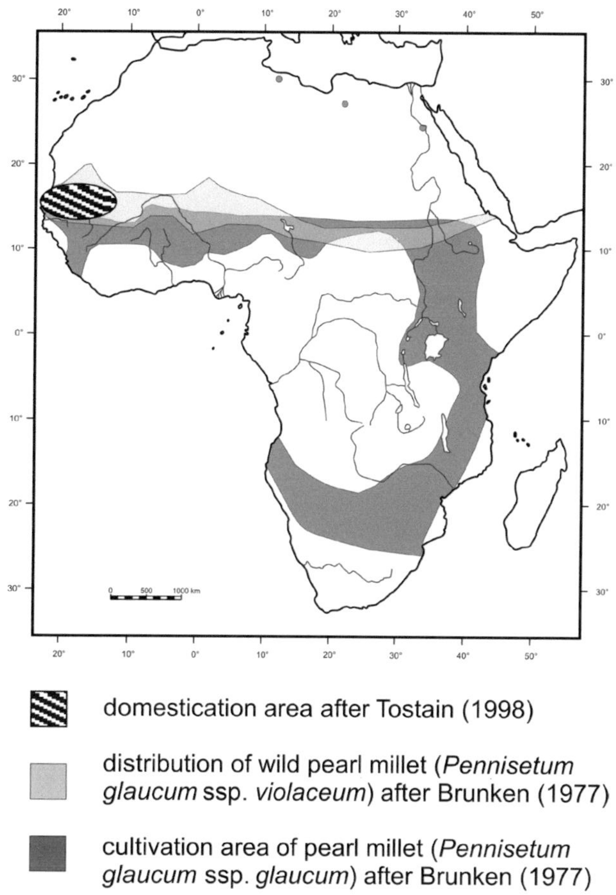

Figure 1. Modern distribution of pearl millet.

Mali, Mauritania, and Ghana. Later on it is also found in Burkina Faso and Nigeria (Fig. 2).

All African archaeobotanical data for pearl millet point to an introduction of the domesticated form, either at the beginning of the occupation or in later phases of the sequences (Kahlheber & Neumann 2007). There are single finds of wild pearl millet (*Pennisetum glaucum* ssp. *violaceum*) in West African sites (compare Kahlheber 2004), but no archaeological site provides convincing evidence for its conscious use. As discussed by Kahlheber & Neumann (2007), the earliest finds of domesticated pearl millet occur in different bio-geographical zones and in varying subsistence systems. The archaeological contexts of the Sahelian sites yielding pearl millet in Mali, Nigeria, and Burkina Faso show a common pattern: full sedentism was not yet established, and plant cultivation probably played a minor role in a mixed economy. Agricultural practice was integrated into existing pastoral or foraging lifeways as a means of minimizing risk and enabling greater predictability of resource exploitation. Sites in Ghana and Mauritania, on the other hand, represent sedentary populations for which pearl millet seems to have

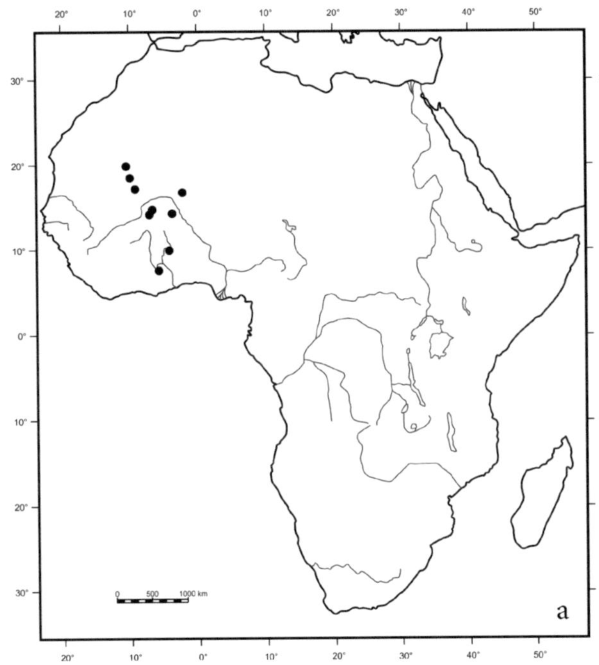

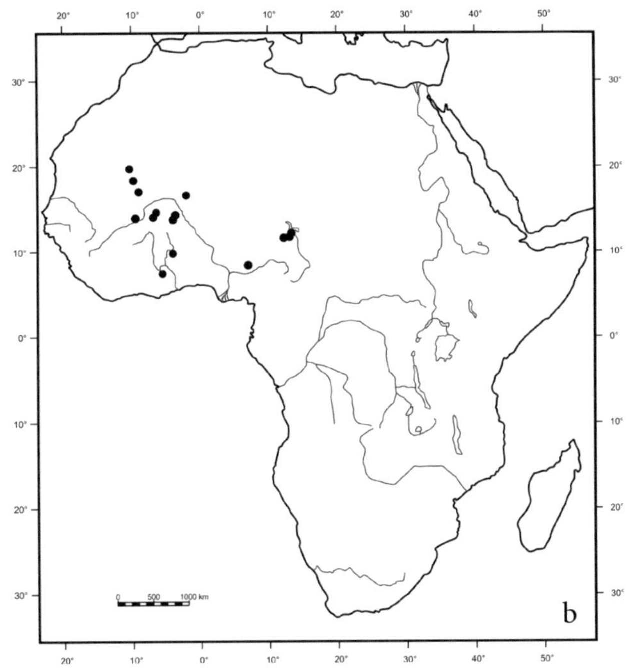

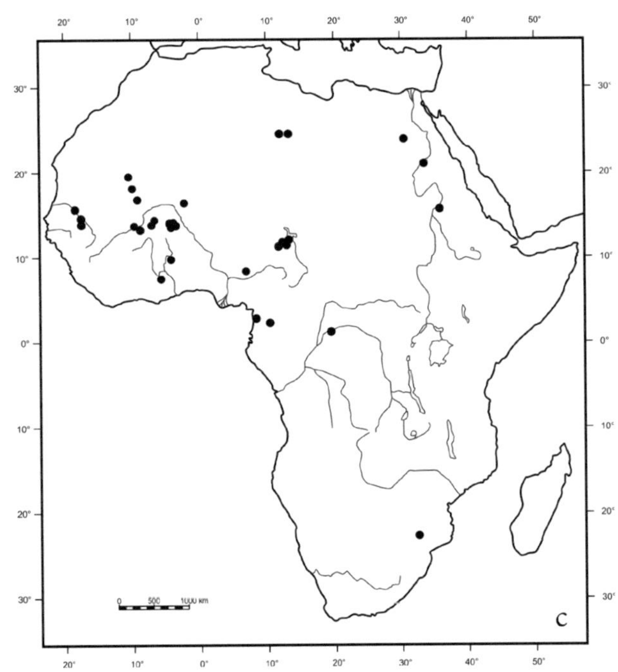

Figure 2. Archaeological sites with finds of domesticated pearl millet dating (a) before 1500 BC; (b) before 500 BC; and (c) before 500 AD (shown is the first regional evidence; the map makes no claim to be complete).

constituted a significant resource. Most early sites display mono-crop assemblages, and it seems that pearl millet was cultivated with comparatively simple techniques.

During the Iron Age (from 500 BC onwards), pearl millet is widely distributed in semi-arid West Africa (Fig. 2c). Several archaeobotanical assemblages illustrate its crucial role in Sudanian and Sahelian agricultural systems, where it constitutes the basic cereal crop and major carbohydrate source. Its cultivation in combination with pulses, various minor crops, and protected wild or semi-domesticated fruit trees resulted in an intensification of food production. This helped to form the economic base for a sedentary lifestyle, demographic growth, increased division of labour, and social stratification, which eventually resulted in the emergence of the famous West African empires.

The dispersal routes of pearl millet become evident when the archaeobotanical evidence is mapped (Fig. 2): radiocarbon dates demonstrate a chronological spread from west to east and from north to south. Around 1200 BC, pearl millet emerges in the Lake Chad region; it becomes the basic carbohydrate crop of rain-fed agrarian systems around 500 BC. Pearl millet finds in India, dating as early as 2500–2300 BC (Fuller 2003; Fuller et al. 2004) demonstrate that this west to east propagation presumably took place much earlier: since the crop is not native to India, it must have been introduced from Africa.

In the last centuries BC, pearl millet is found in Central Africa (Kahlheber et al. 2013, 2014) as well as in the Libyan Fazzan in the Sahara (Pelling 2005). A few centuries later it shows up in the Nile valley and the oases of the Egyptian Western Desert, as discussed below.

The further distribution of pearl millet to eastern and southern Africa is poorly documented. Between 600 and 1000 AD it was the principal grain crop of Swahili communities on Pemba Island, Tanzania (Walshaw 2005), losing its dominant position to Asian rice only thereafter. The first finds of pearl millet in southern Africa are said to date to 270 AD at the Early Iron Age site of Silver Leaves (Klapwijk 1974; Klapwijk & Huffman 1996). The evidence consists of grain impressions in pottery sherds and documentation is rather insufficient. However, the majority of pearl millet finds appears after 600 AD, when farming communities with fully developed agricultural complexes—including sorghum, finger millet, cow pea, and Bambara groundnut—are found all over southern Africa (Robertson 2001).

4. Pearl millet in Egypt

Pearl millet makes its first appearance in Egypt in the second century AD, simultaneously at Kellis in the Dakhleh oasis (Thanheiser et al. 2002) and at Qasr Ibrim in the Nile valley south of Aswan (Clapham & Rowley-Conwy 2007: Table 3), where plenty of desiccated grains and involucra were recovered. Of somewhat older date, possibly Ptolemaic, may be the finds from El Deir in the Kharga oasis. However, the final dating is still pending (Alan Clapham 2014, personal communication). The North Kharga Oasis Survey yielded finds dating to the fourth century AD. Quite interestingly, all of the grains are charred there, while the involucra are desiccated (Alan Clapham 2014, personal communication). More finds come from Late Roman/Early Islamic levels at Hisn al-Bab, a fortress on the river Nile that marked the frontier between Egypt and Nubia (Alan Clapham 2014, personal communication) and from the Islamic Red Sea port at Quseir al-Qadim (Van der Veen 2011: 103).

Additional finds from the Kharga oasis are problematic. At Roman ʻAyn-Manâwir, what appears from the photo to be pearl millet was identified as *Cenchrus* cf. *ciliaris* (Newton et al. 2005: Fig. 10). If material from the Persian level was identified using the same criteria (Agut-Labordère & Newton 2013; Newton et al. 2013), there is a possibility of an earlier introduction of pearl millet in Kharga.

Pearl millet is one of the most frequent and abundant plant remains in domestic areas at Kellis, which was inhabited from the second century AD to c. 360/380 AD. It is mainly parts of the inflorescence, especially fragments of the rachis and involucra, that are present. A storage bin in Area C/2/4 yielded c. 39 kg of such remains mixed only with some flax capsule fragments. Grains, however, are rather rare (Fig. 3). In contrast to the rich finds from Kellis, contemporaneous Amheida in the western part of the oasis (in antiquity it would have been a day trip away from Kellis) yielded only a few involucra (Thanheiser & Walter 2015). This could be the result of taphonomic conditions. While the matrix at Kellis is dry and supports conservation by desiccation, the soil at Amheida is damp. Therefore hardly any desiccated plant remains are present and the dominant form of conservation is by charring. Maybe the fragile bristles of the involucrum did not readily survive contact with fire in any identifiable form. This hypothesis appears to be supported by the fact that, in

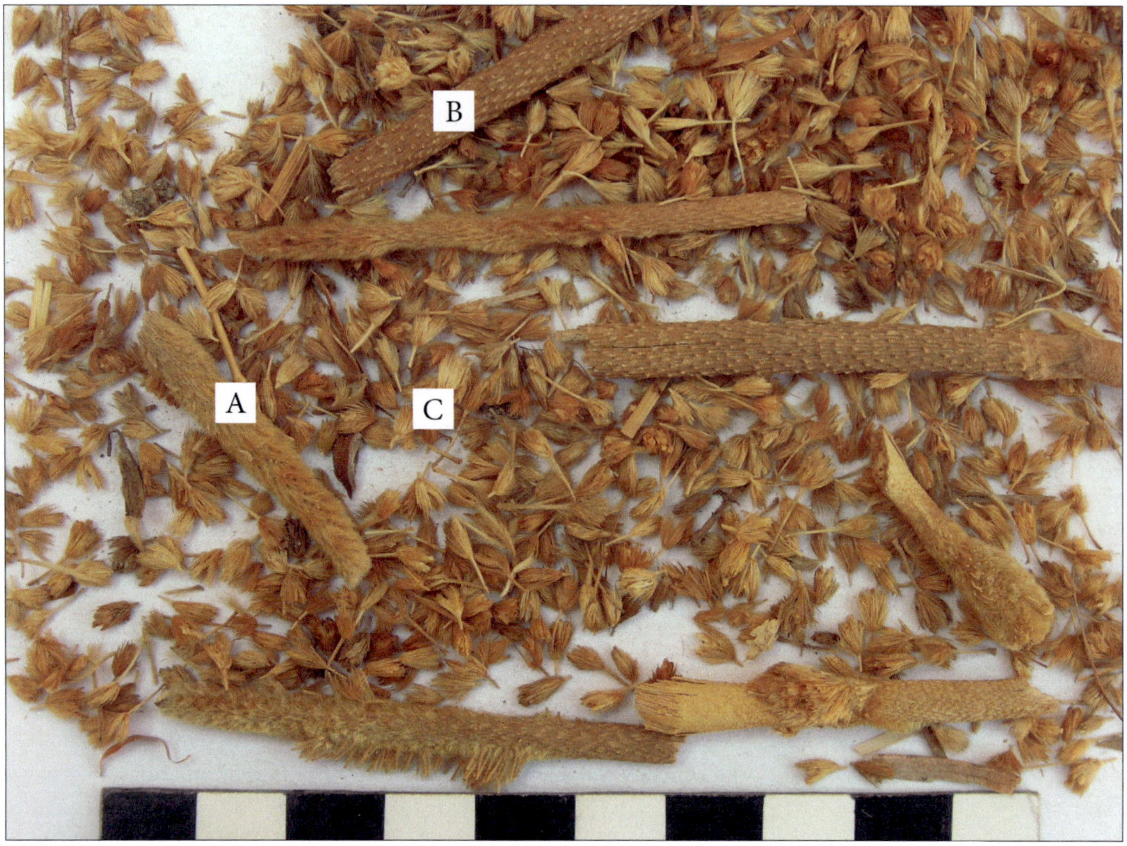

Figure 3. Pearl millet from Kellis, the contents of a storage bin in Area C/2/4. A Fragment of the central axis (rachis) covered with soft hair; B Fragment of the central axis with abscission scars left by the involucra; C Fascicles.

Egypt, so far only sites with desiccated plant remains have yielded appreciable quantities of pearl millet.

Pearl millet, a summer crop, appears in the archaeobotanical record of Egypt in the second century AD. Its presence as one of the major crops at Kellis right from the beginning of the settlement indicates that it was already being cultivated in the area before the settlement was founded. This assumption may eventually be corroborated by possibly earlier finds in the neighbouring Kharga oasis at El Deir. Other important field crops in the Dakhleh oasis are hard wheat (*Triticum durum* Desf.), hulled barley (*Hordeum vulgare* L.), fava bean (*Vicia faba* L.), and flax (*Linum usitatissimum* L.), which are all traditional winter crops. Cotton (*Gossypium* sp.), another summer crop, is present throughout the site (Thanheiser et al. 2002).

Plant cultivation in the Dakhleh oasis depended on artificial irrigation, using ground water fed to the surface along natural vents or through man-made bores. In contrast, the agricultural cycle in the Nile valley was controlled by the yearly flooding of the river. Crops such as emmer wheat (*Triticum dicoccum* Schrank), hulled barley, and flax were sown there after the flood waters had receded, which was usually in early October in the south and four to six weeks later in the north. Harvesting took place about six months later. The land was then left fallow until the next flood, which resulted in a single harvest per year (Murray 2000: 513-515). Experiments with a second annual harvest, however, may date back to the New Kingdom, when a new field type—for which a double tax was levied—appears in the contemporary texts (Schenkel 1978: 59-68).

Pharaonic representations of agricultural sequences sometimes start with the issuing of seed grain, suggesting that this marks the beginning of a new agricultural cycle after the flood waters had receded, i.e. in the autumn. The issuing of seed grain at an unusual time (between January and March) and several contracts dealing with agricultural production lead Schnebel (1925: 145-160) to the conclusion that a second crop was grown in the Fayum in spring during Hellenistic times. The identity of this crop, however, remains unknown.

The introduction of summer crops of African origin, such as pearl millet, marks a profound change in North

African agriculture (Clapham & Rowley-Conwy 2007: 163; Pelling 2005). Cultivated in a second cropping system in addition to the long-established Near Eastern/Mediterranean one, the new African crops imply a diversification of crop inventories and also resulted in an extension of the farming season. Advanced irrigation techniques, such as the *sakkiyeh* and the *tanbur* (water-lifting devices) and the *qanât* or *foggara* (a system of shafts and underground channels for water supply), allowed the dry season cropping of heat-resistant African crops, thus valorising a period of the year previously not used for farming activities. The transition from Near Eastern/Mediterranean winter-grown crops to two cropping seasons is interpreted as resulting in an increase in food production as well as risk-minimization of crop failure and consequently an increase of the carrying capacity of the area.

5. Interpretation of human diet at Kellis using stable carbon isotopes

Stable carbon isotope analysis has been performed on a total of 11 different species of archaeobotanical remains from the site of Kellis (Table 1). The results have revealed significant differences in $\delta^{13}C$ values between plants identified as C_3 and those identified as C_4 (Fig. 4). The only plant identified as a C_4 plant is pearl millet, and its $\delta^{13}C$ value is −10.05‰ (this is an average value based on the

Table 1. Results of stable carbon ($\delta^{13}C$) analyses of archaeological botanical remains from Kellis.

Common name	Scientific name	$\delta^{13}C$ value (‰)
Hard wheat	*Triticum durum*	-22.95
Barley	*Hordeum vulgare*	-23.32
Fava bean	*Vicia faba*	-23.15
Grape	*Vitis vinifera* ssp. *vinifera*	-22.46
Olive	*Olea europaea*	-21.91
Date	*Phoenix dactylifera*	-22.19
Fig	*Ficus carica*	-23.77
Doum palm	*Hyphaene thebaica*	-26.36
Myrtle	*Myrtus communis*	-23.46
Rosemary	*Rosmarinus officinalis*	-23.72
Pearl millet	*Pennisetum glaucum* ssp. *glaucum*	-10.05

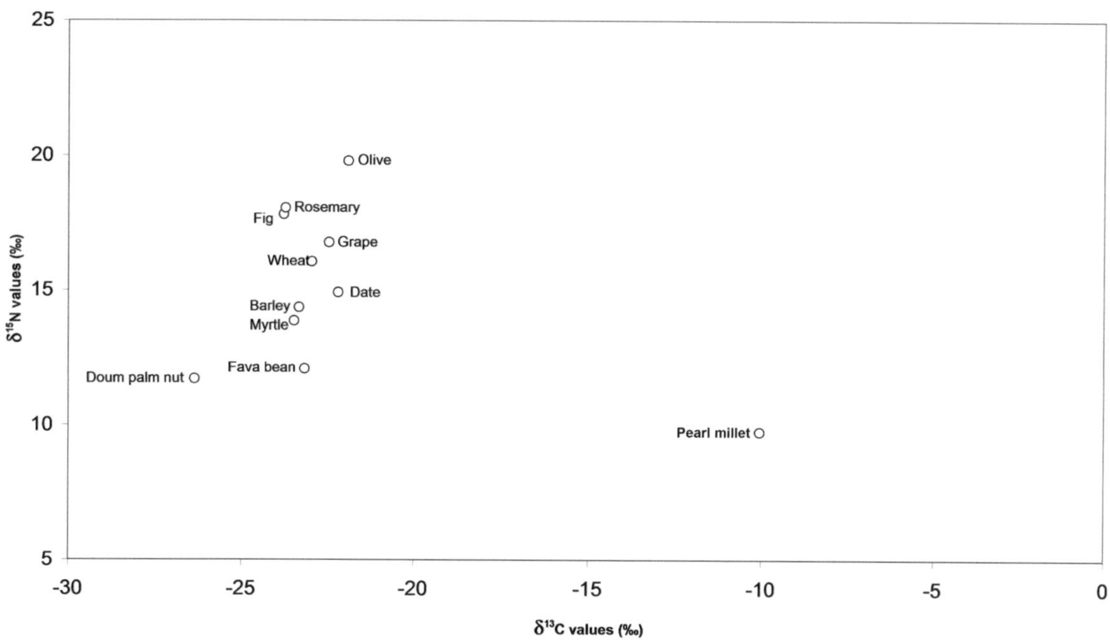

Figure 4. $\delta^{13}C$ and $\delta^{15}N$ results for archaeological botanical remains from Kellis. Although stable nitrogen isotope analyses are not discussed in this paper, they are presented in this graph for ease of display and easier recognition of patterns. Note the differentiation in $\delta^{13}C$ values between the pearl millet and all other plants.

analysis of three different pearl millet samples). All other identified plants belong to the C_3 category, and their $\delta^{13}C$ values (average −23.2‰) reflect this (Table 1; Fig. 4).

Stable carbon isotope patterns established in plants are passed on through the food chain. It is therefore possible to distinguish between diets based on C_3 and on C_4 plants (e.g., Katzenberg et al. 1995). There is approximately a +5‰ difference between plant $\delta^{13}C$ values and those of animal (including human) bone collagen. Thus, humans or other animals consuming a diet based on C_3 plants have bone collagen $\delta^{13}C$ values of approximately −19‰. Humans consuming a diet consisting only of C_4 plants have collagen $\delta^{13}C$ values of approximately −8‰. Individuals who consume a mixed diet of both C_3 and C_4 plants have bone collagen values that are intermediate between the above values.

Stable carbon isotope analyses performed on the animal bones from Kellis have revealed an interesting pattern of plant consumption (Dupras et al. 2001). The bone collagen of a total of six different animal species was analysed for $\delta^{13}C$ values (Table 2). The results revealed that, in comparison to the values for chicken, donkey, pig, and gazelle, the $\delta^{13}C$ values of cows and goats were significantly enriched in $\delta^{13}C$ (Fig. 5). This indicates that a portion of the diet of cows and goats, approximately 20 per cent, would have consisted of C_4 plants, most likely pearl millet (Dupras et al. 2001). The $\delta^{13}C$ values of animal bone collagen would also be reflected in animal products consumed by humans, such as goat's and cow's milk.

Previously published works by Dupras (1999), Dupras and colleagues (2001), Williams (2008), and Williams and colleagues (2011) have demonstrated that the diet

Table 2. Results of stable carbon ($\delta^{13}C$) analyses of archaeological animal remains from Kellis.

Common name	Scientific name	$\delta^{13}C$ value (‰)
Chicken	*Gallus domesticus*	-18.4
Donkey	*Equus asinus*	-18.1
Gazelle	*Gazella gazella*	-17.9
Pig	*Sus scrofa domesticus*	-17.4
Goat	*Capra hircus*	-15.7
Cow	*Bos taurus*	-15.1

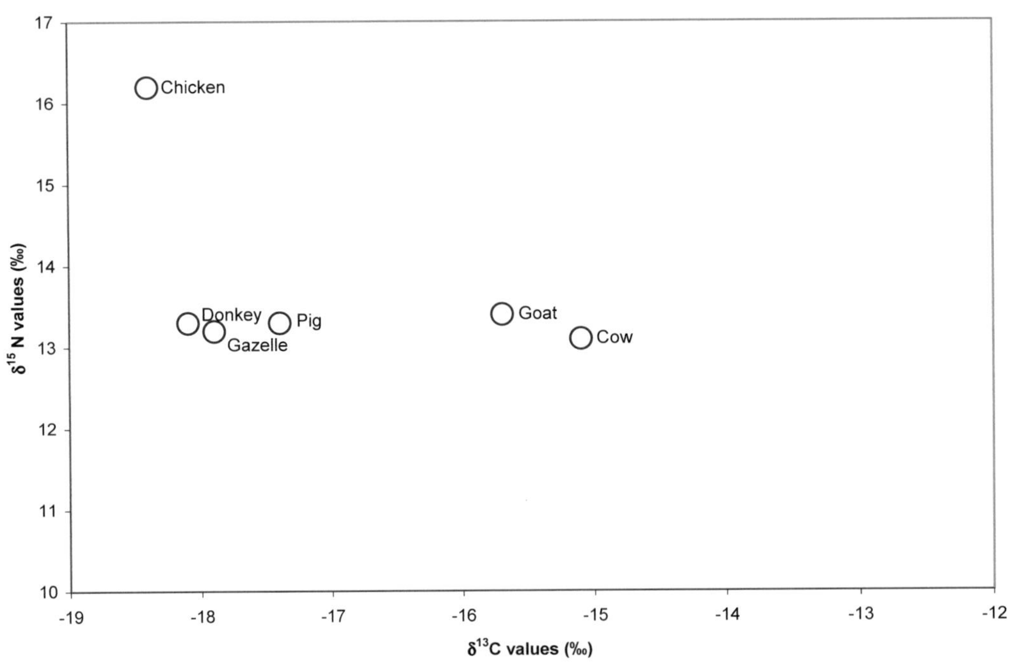

Figure 5. $\delta^{13}C$ and $\delta^{15}N$ results for archaeological animal remains from Kellis. Although stable nitrogen isotope analyses are not discussed in this paper, they are presented in this graph for ease of display and easier recognition of patterns. Note the differentiation in $\delta^{13}C$ values between goat and cow versus the other species.

of the inhabitants of Kellis must have included a food source enriched in ^{13}C. Although stable carbon isotope data reveal that there is some variation in adult diet, the average male and female $\delta^{13}C$ values are –18.8‰ and –19‰, respectively, which indicates a diet heavily reliant on C_3 foods. However, stable carbon data from hair, which illustrates dietary inclusion during the last part of the individual's life, demonstrates the cyclical seasonal use of millet in the diet (Williams 2008; Williams et al. 2011). Stable carbon isotope analysis of the bone collagen of infants and children also reveals the inclusion of a C_4 food source. Between birth and six months of age, infants become increasingly enriched in ^{13}C (an enrichment of approximately 1‰), and after 1.5 years their ^{13}C values slowly decrease until approximately 3.5 years of age, by which age their values are the same as those of adults. The isotope data support the hypothesis that after being exclusively breast-fed for the first six months, infants were introduced to weaning foods that were enriched in ^{13}C (e.g. pearl millet in the form of gruel or cow's/goat's milk from animals that had been fed pearl millet). Stable carbon isotope analysis on tooth dentin from individuals in the Kellis cemetery who had survived the weaning period supports these findings (Dupras & Tocheri 2007). Dentin from teeth forming from birth to three years of age show the same enrichment pattern observed in bone collagen from individuals who died during the weaning process. The results of this longitudinal-type study support earlier findings by Dupras et al. (2001) in which ^{13}C values from bone collagen showed exclusive breastfeeding for six months, followed by an introduction of ^{13}C enriched foods and gradual weaning until three years of age. From the analysis of the botanical and animal remains, it is clear that the food source at the base of the food chain responsible for this enrichment is pearl millet. How the isotope signal was transferred to the infants from Kellis is not known, but it can be surmised that either they were consuming pearl millet directly (through such foods as bread or pap cereals) or they were consuming the by-products of cows and goats (such as milk).

6. Conclusion

With its high drought resistance but good response to irrigation, as well as its requirement of high temperatures for rapid growth, pearl millet would have been the ideal summer crop in the Dakhleh oasis. Here, all water for irrigation came from wells and springs, and the water supply for the crops could therefore be tightly controlled and supplied in critical stages of the plant's growth cycle. In an area where the availability of water and not the presence of arable land is the limiting factor for agricultural production, a drought-resistant summer crop would have considerably enlarged the carrying capacity of the land. As demonstrated by stable carbon isotope analysis, pearl millet not only provided food for human consumption, but also for animals, which would have been particularly important in an area where natural pasture did not exist.

References

Agut-Labordère D. & Newton C., 2013. L'économie végétale à 'Ayn-Manâwir à l'époque perse : archéobotanique et sources démotiques. *Arta* 5, pp. 1-49.

Andrews D.J. & Kumar K.A., 2006. *Pennisetum glaucum* (L.) R.Br. [Internet] Record from PROTA4U. In: Brink M. & Belay G. (eds.), PROTA [Plant Resources of Tropical Africa / Ressources végétales de l'Afrique tropicale], Wageningen. http://www.prota4u.org. Accessed 8 May 2014.

Bagnall R.S. (ed.), 1997. The Kellis Agricultural Account Book (P. Kell. IV Gr. 96) with contributions from C.A. Hope, R.G. Jenkins, A.J. Mills, J.L. Sharpe III, U. Thanheiser & G. Wagner. Dakhleh Oasis Project Monograph 7. Oxbow Monograph 92. Oxbow, Oxford.

Balole T.V. & Legwaila G.M., 2006. *Sorghum bicolor* (L.) Moench. [Internet] Record from PROTA4U. In: Brink M. & Belay G. (eds.), PROTA [Plant Resources of Tropical Africa / Ressources végétales de l'Afrique tropicale], Wageningen. http://www.prota4u.org. Accessed 8 May 2014.

Brunken J., 1977. A systematic study of *Pennisetum* sect. *Pennisetum* (Gramineae). *American Journal of Botany* 64 (2), pp. 161-176.

Brunken J., de Wet J.M.J. & Harlan J.R., 1977. The morphology and domestication of pearl millet. *Economic Botany* 31, pp. 163-174.

Burkill H.M., 1994. *The Useful Plants of West Tropical Africa. Vol. 2. Families E–I.* Royal Botanic Gardens, Kew.

Ceccarelli S. & Grando S., 2006. *Hordeum vulgare* L. [Internet] Record from PROTA4U. In: Brink M. & Belay G. (eds.), PROTA [Plant Resources of Tropical Africa / Ressources végétales de l'Afrique tropicale], Wageningen. http://www.prota4u.org. Accessed 8 May 2014.

Clapham A. & Rowley-Conwy P. 2007. New discoveries at Qasr Ibrim, Lower Nubia. In: Cappers R.T.J. (ed.), *Fields of Change: Progress in African Archaeobotany.* [Proceedings of the 4th International Workshop on African Archaeobotany.] Groningen Archaeological Studies 5. Barkhuis & Groningen University Library, Groningen, pp. 157-164.

Dupras T.L., 1999. Dining in the Dakhleh Oasis, Egypt: Determination of Diet using Documents and Stable Isotope

Analysis. Unpublished Ph.D. Thesis, McMaster University, Hamilton [Canada].

Dupras T.L., Schwarcz H.P. & Fairgrieve S.I., 2001. Infant feeding and weaning practices in Roman Egypt. *American Journal of Physical Anthropology* 115, pp. 204-211.

Dupras T.L. & Tocheri M.W., 2007. An investigation of infant weaning and feeding patterns at Roman period Kellis, Egypt using stable isotope analysis of dental enamel and dentin. *American Journal of Physical Anthropology* 134 (1), pp. 63-74.

FAO. n.d. http://www.fao.org/ag/AGP/AGPC/doc/Gbase/DATA/PF000297.htm Accessed 8 May 2014.

Fuller D.Q., 2003. African crops in prehistoric South Asia: A critical review. In: Neumann K., Butler A. & Kahlheber S. (eds.), *Food, Fuel and Fields: Progress in African Archaeobotany.* [Proceedings of the 3rd International Workshop on African Archaeobotany.] Africa Praehistorica 15. Heinrich-Barth-Institut, Köln, pp. 239-271.

Fuller D.Q., Korisettar R., Venkatasubbaiah P.C. & Jones M.K., 2004. Early plant domestications in southern India: Some preliminary archaeobotanical results. *Vegetation History and Archaeobotany* 13, pp. 115-129.

Harlan J.R., 1971. Agricultural origins: Centers and non-centers. *Science* 174, 468-474.

Kahlheber S., 2004. Perlhirse und Baobab—Archäobotanische Untersuchungen im Norden Burkina Fasos. Unpublished Ph.D. thesis, J.W. Goethe-Universität, Frankfurt, http://publikationen.ub.uni-frankfurt.de/volltexte/2005/561/

Kahlheber S., Eggert M.K.H., Seidensticker D. & Wotzka H.-P., 2014. Pearl millet and other plant remains from the Early Iron Age site of Boso-Njafo (Inner Congo Basin, Democratic Republic of the Congo). *African Archaeological Review* 31, pp. 479-512.

Kahlheber S., Höhn A. & Neumann K., 2013. Plant and land use in southern Cameroon 400 B.C.E.–400 C.E. In: Stevens C.J., Nixon S., Murray M.A. & Fuller D.Q. (eds.), *Archaeology of African Plant Use.* UCL Institute of Archaeology Publications 61. Left Coast Press, Walnut Creek, pp. 113-128.

Kahlheber S., Höhn A. & Rupp N., 2009. Archaeobotanical studies at Nok sites: An interim report. *Nyame Akuma* 71, pp. 2-15.

Kahlheber S. & Neumann K., 2007. The development of plant cultivation in semi-arid West Africa. In: Denham T.P., Iriarte J. & Vrydaghs L. (eds.), *Rethinking Agriculture: Archaeological and Ethnoarchaeological Perspectives.* One World Archaeology. Left Coast Press, Walnut Creek, pp. 320-346.

Katzenberg M.A., Schwarcz H.P., Knyf M. & Melbye F.J., 1995. Stable isotope evidence for maize horticulture and paleodiet in southern Ontario, Canada. *American Antiquity* 60 (2), pp. 335-350.

Klapwijk M., 1974. A preliminary report on pottery from the north-eastern Transvaal, South Africa. *South African Archaeological Bulletin* 19, pp. 19-23.

Klapwijk M. & Huffman T.N., 1996. Excavations at Silver Leaves: A preliminary report. *South African Archaeological Bulletin* 51, pp. 84-93.

Mangat B.K., Maiti R.K. & Khairwal I.S., 1999. Pearl millet biology. In: Khairwal I.S., Rai K.N., Andrews D.J. & Harinarayana G. (eds.), *Pearl Millet Breeding.* Science Publishers, Enfield, pp. 1-27.

Manning K., Pelling R., Higham T., Schwenniger J.-L. & Fuller D.Q., 2011. 4500-year old domesticated pearl millet (*Pennisetum glaucum*) from the Tilemsi Valley, Mali: New insights into an alternative cereal domestication pathway. *Journal of Archaeological Science* 38 (2), pp. 312-322.

Murray M.A., 2000. Cereal production and processing. In: Nicholson P.T. & Shaw I. (eds.), *Ancient Egyptian Materials and Technology.* Cambridge University Press, Cambridge, pp. 505-536.

National Research Council. 1996. *Lost Crops of Africa. Vol. 1. Grains.* National Academy Press, Washington.

Newton C., Gonon T. & Wuttmann M., 2005. Un jardin d'oasis d'époque romaine à ʻAyn-Manâwir (Kharga, Égypte). BIFAO 105, pp. 167-195.

Newton C., Whitbread T., Agut-Labordère D. & Wuttmann M., 2013. L'agriculture oasienne à l'époque perse dans le sud de l'oasis de Kharga (Égypte, V^e–IVe s. AEC). *Revue d'Ethnoécologie* 4, pp. 1-20.

Oyen L.P.A. & Andrews D.J., 1996. *Pennisetum glaucum* (L.) R. Br. In: Grubben G.J. & Partohardjono S. (eds.), *Plant Resources of South-East Asia. Vol. 10. Cereals.* Backhuis, Leiden, pp. 119-123.

Pelling R., 2005. Garamantian agriculture and its significance in a wider North African context: The evidence of the plant remains from the Fazzan project. *Journal of North African Studies* 10 (3–4), pp. 397-411.

Robertson J.H., 2001. South African Early Iron Age in Zambia. In: Peregrine P.N. & Ember M. (eds.), *Encyclopedia of Prehistory. Vol. 1. Africa.* Kluwer Academic/Plenum Publishers, New York, pp. 260-271.

Schenkel W., 1978. Die Bewässerungsrevolution im Alten Ägypten. Verlag Philipp von Zabern, Mainz.

Schnebel M., 1925. Die Landwirtschaft im hellenistischen Ägypten. C. H. Beck Verlagsbuchhandlung, München.

Singh F. & Nainawatee H.S., 1999. Grain quality traits. In: Khairwal I.S., Rai K.N., Andrews D.J. & Harinarayana G. (eds.), *Pearl Millet Breeding.* Science Publishers, Enfield, pp. 157-183.

Stoskopf N.C., 1985. *Cereal Grain Crops.* Reston Publishing Company, Reston.

Thanheiser U. & Walter J., 2015. Plant use in a Romano-Egyptian household in the third century CE. In: Boozer A.L., Amheida II: A late Romano-Egyptian House in the Dakhleh Oasis: Amheida House B2. The Institute for the Study of the Ancient World, New York, pp. 375-392. http://dlib.nyu.edu/awdl/isaw/amheida-ii-house-b2/.

Thanheiser U., Walter J. & Hope C.A., 2002. Roman agriculture and gardening in Egypt as seen from Kellis. In: Hope C.A. & Bowen G.E. (eds.), *Dakhleh Oasis Project: Preliminary Reports on the 1994–1995 to 1998–1999 Field Seasons*. Dakhleh Oasis Project Monograph 11. Oxbow, Oxford, pp. 299-310.

Tostain S., 1998. Le mil, une longue histoire: hypothèses sur sa domestication et ses migrations. In: Chastanet M. (ed.), *Plantes et paysages d'Afrique : une histoire à explorer*. Karthala, Paris, pp. 461-490.

Van der Veen M. (ed.), 2011. *Consumption, Trade and Innovation: Exploring the Botanical Remains from the Roman and Islamic Ports at Quseir al-Qadim, Egypt*. Journal of African Archaeology Monograph Series 6. Africa Magna Verlag, Frankfurt.

Walshaw S.C., 2005. Swahili urbanization, trade and food production: Archaeobotanical perspective from Pemba Island, Tanzania, AD 600–1500. Unpublished Ph.D. Thesis, Washington University, St. Louis.

Williams L.J., 2008. Investigating Seasonality of Death at Kellis 2 Cemetery using Solar Alignment and Isotopic Analysis of Mummified Tissues. Unpublished Ph.D. Thesis, University of Western Ontario, London.

Williams L.J., White, C.D., Longstaffe, F.J., 2011. Improving stable isotopic interpretations made from human hair through reduction of growth cycle error. *American Journal of Physical Anthropology* 145, pp. 125-136.

Wood use in Graeco-Roman Karanis (Fayum, Egypt): Local origin or import? First results

Caroline Vermeeren

BIAX Consult
Hogendijk 134
1506 AL Zaandam, the Netherlands

E-mail: vermeeren@biax.nl

Abstract. Until recently, excavations in Graeco-Roman Karanis (Fayum Oasis, Egypt) have paid little attention to archaeobotanical research, in spite of the agricultural importance of the city. A new research project, focusing on the introduction and large-scale organization of agriculture during the Graeco-Roman period, provided an opportunity to look at desiccated wood from the site. The main question was whether the wood used for objects and construction was obtained locally or imported—the latter implying a completely different organization.

The wood constructions were dominated by small, curved *Tamarix, Olea,* and *Acacia* branches, with some other taxa. Rare finds of larger wood constructions that had survived earlier demolitions, such as beams and steps, again showed these three taxa, but were dominated by *Phoenix*. Using the results of the analysis of woodworking remains, non-worked wood, charcoal, and the modern vegetation, it is argued that these taxa were locally obtained. The same taxa were used to make the wooden objects, but imported taxa (*Abies*, *Alnus*, *Buxus*, *Cupressus*, *Fagus*, *Fraxinus*, and *Phillyrea/Rhamnus*), probably obtained from the Mediterranean, were also recorded. *Salix* was also used for objects; it is not yet clear, however, whether it was imported or of local origin.

Keywords: Egypt, wood use, Graeco-Roman, imported wood, local wood

1. Introduction

The Graeco-Roman village of Karanis was situated in the Fayum Oasis, receiving water from the Nile by means of a drainage canal. It was founded in the third century BC and abandoned at the end of the fifth century AD. In the nineteenth and twentieth centuries, several excavations were carried out (see Boak & Peterson 1931; Husselman 1979), but little attention was given to archaeobotanical research. Also, many of the waste layers (*sebakh*) were levelled and used as fertilizer on the (surrounding) fields, which have poor, sandy soils. Larger wooden structures were demolished, the wood reused in other buildings or burned as fuel. In this way, important information was lost.

The many storage rooms, (olive) presses, and millstones that were uncovered indicate that agriculture played an important role in this rural village. Therefore the main purpose of a new research project in Karanis was to focus on agriculture and the associated organization during the Graeco-Roman period. The project was started in 2003 by René Cappers (Rijksuniversiteit Groningen) and Willeke Wendrich (University of California, Los Angeles). Many archaeobotanical results were obtained from middens, ash layers, plaster, mortar, and clay tiles (see also Fantone, this volume). Some charcoal was analysed by John Marston (2007).

During the 2009 and 2010 excavations, research on desiccated wood from the Roman period[1] was initiated by the author. Many wooden objects and construction wood could be investigated before they found their way

1 More precise dates are not available at this moment; it is anticipated that adjusted dates will be available for upcoming publications by the author in volumes edited by Cappers and Wendrich, respectively (Vermeeren in prep. a, b).

to the museum (Vermeeren in prep. a, b). It was also possible to look at the remaining wood structures *in situ* in the many standing buildings at the excavation site. This was a unique and possibly last opportunity, because the wooden structures deteriorate and decay quickly.

The aim of this research was to find out more about wood use in Karanis: Which taxa were used? Which woodworking methods were performed and with what kind of tools? Are there any woodworking remains to show that work took place on the site? Together the results will be used to answer the question whether the wood was obtained locally or imported. For this, it was also necessary to get information about non-worked wood and charcoal used in Karanis.

The selection of wood depends on three factors: availability, quality, and symbolic value (Asensi Amoros & Détienne 2008; Lange 2009; Vermeeren & Brinkkemper 2005). If no special properties are needed, only available wood is used. Here are just three of the many examples that exist. In firewood collection strategies during the Neolithisation in the Netherlands (Out 2010), the main factor was availability. In Iron age Vlaardingen (the Netherlands), houses and trackways were mainly constructed of *Alnus* (alder), the locally abundant tree according to the pollen and plant macroremain research (Hänninen 2011); research on management activities found no evidence for this (Vermeeren 2011). In a twelfth-century farm in Gouda (the Netherlands), wood and pollen were mainly of *Alnus* and *Fraxinus excelsior* L. (ash), together with small amounts of all the other taxa from the local eutrophic peatland (Bakels et al. 2000). Only one feature is an exception, as will be discussed below.

If available wood sources are not sufficient enough for a particular purpose, the second factor, quality, becomes an issue: either the taxon with the best quality within the available woody vegetation is chosen, or quality is improved through management methods, or a taxon of better quality is imported. Examples of the first strategy are found in many European Mesolithic excavations, where there is a strong correlation between artefact types with specific demands and wood taxa with specific qualities. Bows are made of *Taxus* (yew) and *Ulmus* (elm); paddles, of *Fraxinus excelsior*; fish traps, of *Cornus sanguinea* L. (dogwood) and *Salix* (willow); ropes, of *Tilia* (lime) and *Fraxinus excelsior/Ulmus*; and canoes, of *Quercus* (oak) or *Tilia*, despite these being sometimes rare (or at least not dominant) taxa in the vegetation (Louwe Kooijmans, Hänninen et al. 2001; Louwe Kooijmans, Vermeeren et al. 2001; Out 2009: 229-254). This correlation is still common in modern wood use (Kalkman et al. 2003: 245-251; Wiselius 1990).

Suggestions that management took place to improve the quality of timber used in trackways are made by Casparie (2005), and even stronger implications are found in his work on long, thin *Cornus sanguinea* withies in fish traps presented by Out (2009: 229-254). The use of long-distance trade to import a taxon of better quality was common in the Roman period. It can be seen in special objects—such as a small table of Indian *Tectona* (teak) wood in the Roman harbour of Berenike (Egypt; Vermeeren 2000/2003) and the majority of the funeral portraits from the Fayum area, which were made of imported *Cedrus* (cedar), *Fagus* (beech), and *Tilia* (lime) (Asensi Amoros & Détienne 2008)—but also in larger constructions, such as a Roman road in Valkenburg (the Netherlands), which, according to the dendrochronological data, indicates the use of *Quercus* wood originating from Germany (Bult et al. 1989).

The third factor, symbolic value, is very much culturally fixed and difficult to prove. It can be a preference for specific taxa based on religious reasons, such as the use of evergreen coniferous Christmas trees, but also on other, now-unknown reasons or taboos. Examples are found in the above-mentioned medieval farm in Gouda, which was built of available (local) wood with the exception of a triangular feature near the entrance, made of split branches of *Pinus* (pine) and *Abies alba* Mill. (silver fir), taxa that were not growing anywhere near the site during that period. The material had not been reused. Symbolic finds are also mentioned by Lange (2009) and Therkorn et al. (2009: 148-152), among others.

Because of the mixture of these three factors, the choice of wood is not always a direct reflection of the local vegetation. Taxa may have been imported and locally available taxa ignored because higher-quality wood was required or because there was a cultural taboo on using them. Another problem is that what is found in an excavation is only a part of the wood originally used, because of taphonomic processes and earlier demolitions.

All this makes it difficult to find out which taxa in the dataset are local in origin and which are imported. Pollen data, which are normally used to get an idea of the vegetation in the surrounding landscape, are not available in the dry environment of Karanis. In an attempt to differentiate local versus import, the wood was divided into four categories: (1) objects, (2) construction wood, (3) woodworking remains or waste material, and (4) non-worked wood (these categories are also used in Hänninen & Vermeeren [1998, in prep.]). Although some overlap between the categories cannot be avoided and boundaries are not always completely strict, this division makes it possible to compare the wood results of the different categories. Charcoal results give additional information here. The same three factors (availability, quality, and symbolic value) play a role in the choice of firewood. There are some examples of taxa selected for specific burning capacities, such as Pomoidea[2] in two Dutch sites (Hänninen & Vermeeren 1995; Vermeeren & de Hingh 2000). Ethnographic observations in Egypt (Vermeeren & Cappers 1997/2002) show that 'cultural rules' play a role in the preference for certain taxa as firewood and the avoidance of others, a choice that can only partly be correlated with the quality of the taxa. More often the results from charcoal research show which taxa were available around the site (e.g. Hänninen & Vermeeren 1998, in prep.; Out 2010).

The idea is that if a taxon is present in all categories, it must have been locally available. It may possibly still play an (important?) role in the modern vegetation. If a taxon is imported, that will be for special quality or symbolic demands, and therefore it will be present in more special finds, such as objects or special constructions (Category 1 and Category 2). Small amounts of woodworking remains (Category 3) can be expected if imported beams or planks are finished at the site (fitting to shape, planing, sharpening of points), but this waste material will be scarce. The tree may since have become a part of the modern-day vegetation, but it was not growing there locally during the Roman period, so it is not expected to be present in the non-worked wood (Category 4) and not or seldom in the charcoal.[3] It is plausible to assume that climatic conditions in Egypt in Roman times were not much different from the modern situation—at least not significantly influencing the composition and limits of woody vegetation. So the modern-day vegetation descriptions (Boulos 1999, 2000, 2003, 2005; Täckholm 1974; Täckholm et al. 1941) can be used as a basis. We do, however, have to take into account the modern introductions of the past two centuries.

Results of categories 1 (objects) and 2 (construction wood) will be presented here, together with the preliminary results of categories 3 (woodworking remains) and 4 (non-worked wood) and the charcoal results of 2007. The complete results will be published as a chapter in the book on the archaeobotany of Karanis (Vermeeren in prep. a).

2 It probably concerns hawthorn, but based on wood anatomy this cannot be distinguished from apple and pear.

3 Imported wood may become part of the charcoal category if woodworking remains, discarded constructions, or broken objects are burned.

2. Methods

Numerical measurements were recorded, and a standardized code was assigned that indicates which part of the tree was used.[4] Traces of woodworking methods (tool marks) as well as user marks were described and, if in good condition, photographed. Finally, the wood was identified to the lowest taxonomic level possible. Because the objects and the well-preserved construction wood have to be handled with the greatest care, only microscopically small samples (thin-sections) were taken under an incident light microscope at 40× magnification. For identification it is necessary to cut wood on three different planes: transverse, radial, and tangential. The slides are soaked in soapy water to remove most of the air and subsequently sometimes bleached before they are embedded in corn syrup. After drying, the slides are examined under a transmitted light microscope (magnification up to 250×) to obtain information on the anatomical features. Because the state of preservation of the desiccated wood from this excavation is mostly quite poor, it is sometimes almost impossible to make thin-sections. In that case the transverse section, the most difficult to make on brittle desiccated wood, was only examined under the incident light binocular to describe the most important features. Most wood can be identified to the genus level, following Fahn et al. (1986), Neumann (1989), Neumann et al. (2000), Schweingruber (1978, 1990), Wheeler et al. (1989), and the website InsideWood.

Not all the construction wood could be taken from the field; many timbers had to be investigated *in situ* because they were still part of structures. These pieces were described in the field and sampled in order to obtain a taxonomic identification. Their exact position was plotted with GPS.

3. Results

The desiccated wood that was collected during the two field seasons was divided into the four categories, and estimations were made of the total numbers for the 2009 season. For 2010, the proportions of the four categories were more or less the same as for 2009, and together there are more than 15 000 pieces. Categories 1 (objects, N=82) and 2 (construction wood, N=326) make up only between 2 and 3 per cent of the total. These are almost all

4 A form with these standardized codes can be obtained from the author.

described and identified. Category 3, the woodworking remains, represent less than 10 per cent of the total; only a few pieces have been investigated so far (N=17). For Category 4, the non-worked wood, which is by far the most numerous group, almost 90 per cent, only preliminary results are available.

3.1 Category 1: Objects

The objects include many domestic and small craft elements, such as bowls, plates, cups, boxes, stoppers, combs, and spindles, but also ornaments of unknown origin and small worked objects that were part of small constructions of unknown purpose (Vermeeren in prep. a).

This category exhibits the largest variation in taxa (Fig. 1). Of the 11 taxa identified, *Olea* (olive) and *Tamarix* (tamarisk) dominate. Only *Tamarix* and one *Salix* species (willow) are part of the modern local vegetation. The other taxa, *Abies* (fir), *Alnus* (alder), *Buxus* (boxwood), *Cupressus* (cypress), *Fagus* (beech), *Fraxinus* (ash), and *Phillyrea/Rhamnus* (phillyrea/buckthorn), do not occur in the local vegetation, and there are no indications that they did in the past. *Olea* is not an element in the wild vegetation either but was introduced and cultivated in Egypt, probably in the first millennium BC (Newton et al. 2006; Zohary & Hopf 1994: 137-143).

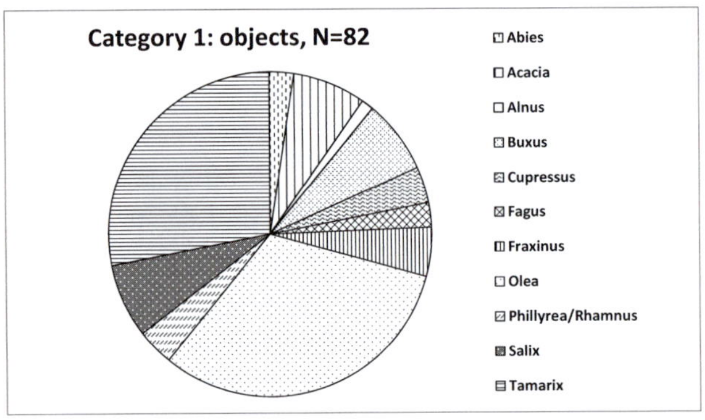

Figure 1. Diagram showing proportions of taxa used for objects.

One of the special finds made of olive wood is a lice comb. In the Roman period, this type of comb is normally made from wood of the strong and beautiful Mediterranean *Buxus* (Derks & Vos 2010). The Romans brought these combs with their personal belongings, from the Mediterranean to every corner of the empire. Among the exceptionally few that are made from other wood taxa is the *Olea* comb from Karanis. This was probably locally manufactured on locally available wood. *Olea* has comparable qualities to *Buxus*. All the other combs from Karanis are made of *Buxus* (Fig. 2).

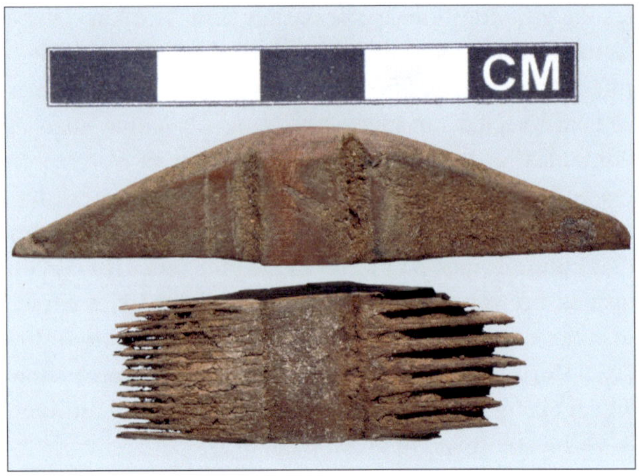

Figure 2. Two fragments of a (lice) comb made of *Buxus* (boxwood). (© Fayum team.)

3.2 Category 2: Construction wood

There was a variety of construction elements in Karanis. Smaller elements, such as pins or broken beams and planks, were taken from the field and described in the research house, whereas larger elements, such as doorpost and steps, or elements that were incorporated into the walls and ceilings and could not be dismantled, were investigated at the site *in situ*. The wall elements were often small, curved *Tamarix, Olea,* and *Acacia* branches (Fig. 3) that seem to have functioned as a kind of wall reinforcement.

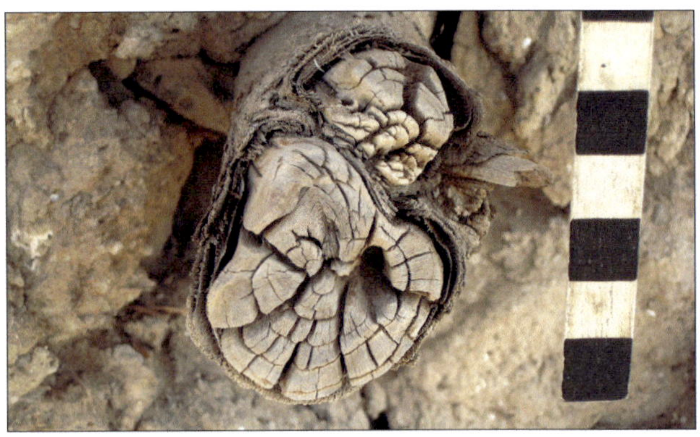

Figure 3. Branches of *Tamarix* used in the ceiling of a granary.

The dominant taxa are the same as in Category 1: *Tamarix* and *Olea* (Fig. 4). A close third is *Acacia*, and a new taxon is *Phoenix* (date palm). Wild ancestors of this palm grow in Egypt. Cultivars have been present since the sec-

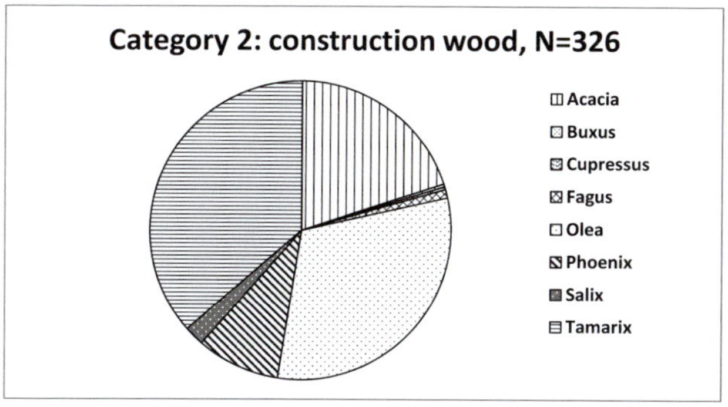

Figure 4. Diagram showing proportions of taxa used as construction wood.

ond millennium BC (Zohary & Hopf 1994: 157-162). It is cultivated in the oasis today, together with *Olea* and other crops. Several species of *Acacia* occur in Egypt, but it is not found in the modern vegetation in the Fayum today. Other taxa found in the construction wood category are *Buxus*, *Cupressus*, *Fagus,* and *Salix.*

There is considerable difference in size within this group. The taxa of the many small elements, such as pins and branches, numerically dominate the taxa of the few remains that were left of the larger constructions (as these were the first to be taken out during the ancient demolitions). To see if there is an effect on the distribution of taxa, the larger construction elements made of trees of more considerable size (diameter >10 cm, N=42) were separated from the total (Fig. 5). In this subsample, the percentages of the three dominant taxa *Tamarix*, *Olea,* and *Acacia,* drop slightly and *Buxus*, *Cupressus*, *Fagus,* and *Salix* disappear completely—all in favour of *Phoenix,* which becomes the dominant element. The palm is cultivated for its fruits, and the leaves are used as roofing material and in small constructions. It forms a long, straight, strong trunk, consisting of fibres, that works well in larger constructions. Some large trunks were used as steps (Fig. 6) and beams.

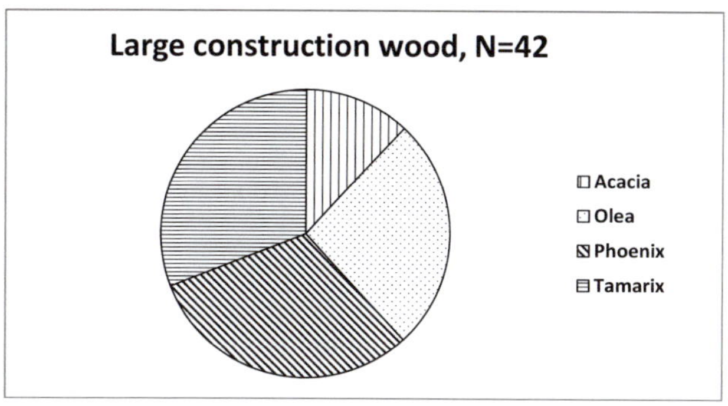

Figure 5. Diagram showing proportions of taxa used in larger constructions.

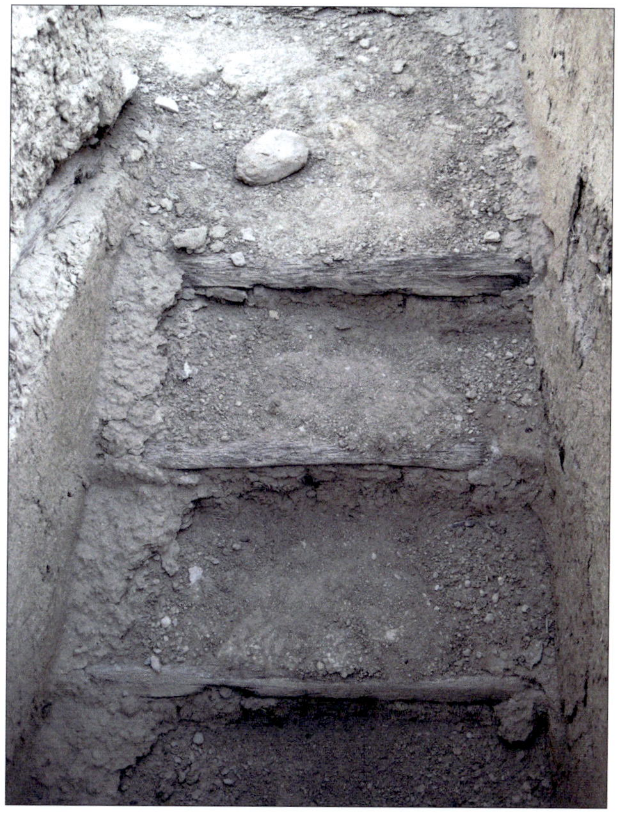

Figure 6. Steps made of halved *Phoenix* trunks.

The preservation of the construction wood ranges from very poor to almost excellent. On the better-preserved pieces, traces of various woodworking methods were found, such as sawing (Fig. 7), planing, gouging, drilling, chopping, turning, and glueing. User marks could also sometimes be seen on the wood (knife cuts, friction marks, etc.). More detail on woodworking methods will be given in Vermeeren (in prep. a).

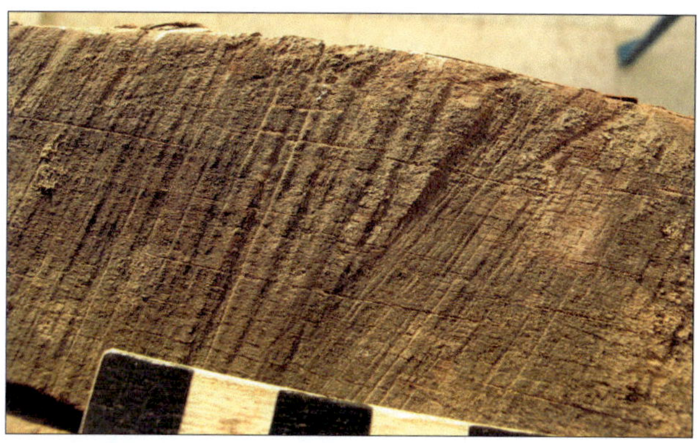

Figure 7. Saw marks on a beam.

3.3 Category 3: Woodworking remains

The number of investigated pieces is too small to draw any definite conclusions here, but among the few pieces examined (N=17), the same two taxa dominate: *Olea* and *Tamarix* (Fig. 8). They are found as cut-offs of branches, knots, and woodchips with bark. There is also some waste material of *Acacia* and *Abies*. Additional taxa will probably be found as more material is examined. Some examples of woodworking remains are given in Figure 9.

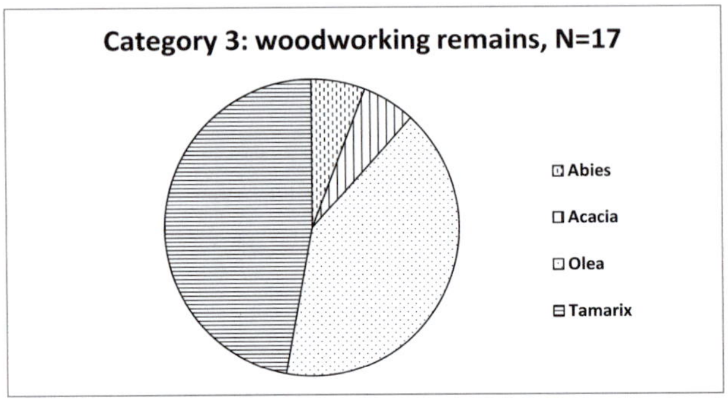

Figure 8. Diagram showing proportions of taxa found as woodworking remains.

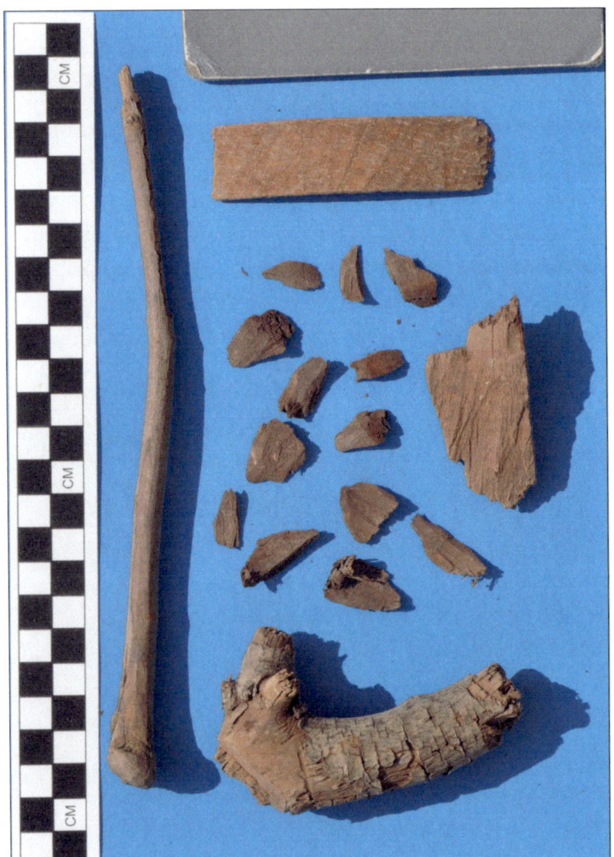

Figure 9. Example of woodworking remains (category 3): woodchips, cut-offs, and waste material. (© Fayum team.)

3.4 Category 4: Non-worked wood

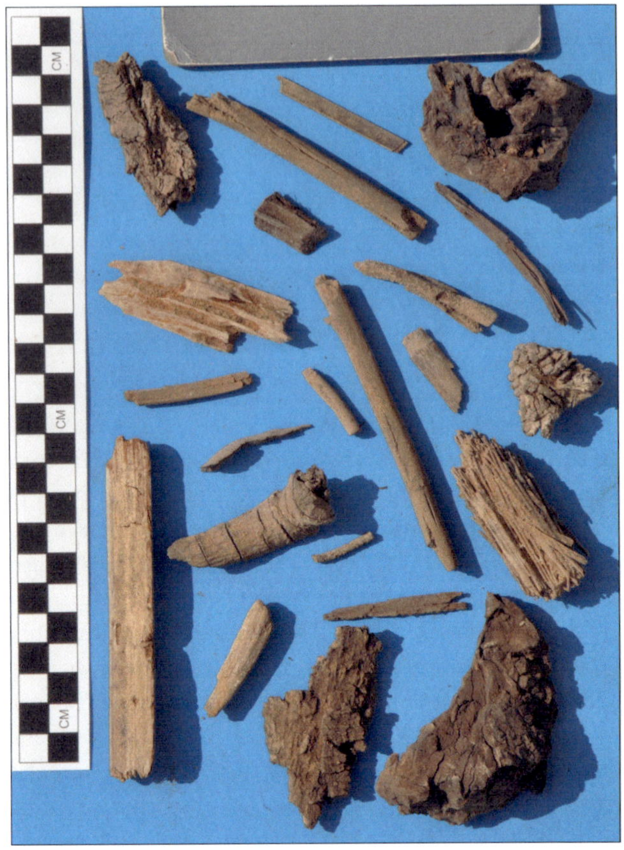

Figure 10. Example of non-worked wood (category 4): branches, fragments, knots. (© Fayum team.)

The first impression is that a majority of the collected non-worked wood is not strictly speaking wood, although it may sometimes originate from trees. In addition to flower heads and leaves of *Tamarix, Phoenix,* and *Acacia,* large amounts of woody stems probably belonging to Poaceae (grasses) were present. These macroremains are an indication for locality; the results of the non-worked wood analysis will be combined with the results of other macroremain analyses from the site in forthcoming research. Some examples of non-worked wood can be seen in Figure 10. The same three taxa (*Tamarix, Phoenix,* and *Acacia)* were recognized in the pieces of bark and the branches. *Olea* is probably present too, but this needs to be confirmed.

3.5 Charcoal

The charcoal research from 2007 (Marston 2007) is dominated by *Tamarix* and *Olea,* followed by *Acacia* and *Phoenix* (Fig. 11). Small numbers of remains of *Salix,*

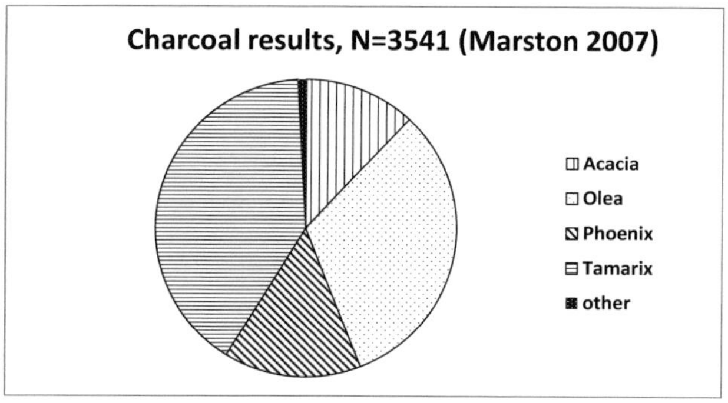

Figure 11. Diagram showing proportions of taxa found as charcoal, by weight (data from Marston 2007).

Pinus, cf. *Fagus*, and *Juniperus* (juniper) were also identified. The total number of identifications was over 3541, with a weight of 1044.7 g. The proportions in Figure 11 are based on weight percentages.

Acacia provides very good fuel (Vermeeren 1998; Vermeeren & Cappers 1997/2002), and the same is true for *Olea* (Oduor et al. 2012). The other two taxa were probably mainly used because they were abundantly available in and around the Roman city, for these taxa do not provide particularly good firewood. A comparable phenomenon was seen in the charcoal results of several middens near the Roman harbour of Berenike, on the east coast of Egypt, which where dominated by the mangrove *Avicennia* (Vermeeren 1998, 1999). The Romans probably used it because it became abundantly present when they cleared the coastline for the extension of the harbour (Vermeeren 2000/2003; Vermeeren & Cappers 1997/2002).

4. Conclusions and discussion

Although the research is still ongoing and data from woodworking remains are scarce and those from the non-worked wood are only preliminary, some conclusions can be drawn. The main taxa used in Roman Karanis are *Tamarix, Acacia, Olea,* and *Phoenix.* Their proportions differ among the four categories of wood (objects, construction wood, woodworking remains, and non-worked wood).

Tamarix and *Acacia* are present in all the wood categories and also in the charcoal and macroremains. It is assumed that a taxon present in all categories must have been locally available. *Tamarix* is and was an important woody element of the vegetation, used for everything in daily life with low quality demands, including firewood. *Acacia* wood is strong and makes very good fuel. Its proportion in the construction wood was higher than its proportion in the charcoal, so its qualities were probably even more valued for construction than for firewood. In the modern-day vegetation it has not been found so far, although it can and did grow in the Fayum in the past. It is possible that *Acacia* vanished due to overexploitation. *Olea* wood is found as objects, construction material, woodworking remains, and charcoal. It has not yet been found in the non-worked wood category, but that is probably only because the number of pieces identified is still very low; the composition could change when identification of this category has been completed. Its occurrence in all categories means the tree was locally available. However, the tree is not native to Egypt. It was introduced, probably as early as the first millennium BC (Newton et al. 2006; Zohary & Hopf 1994). It seems to have been cultivated in Egypt ever since, and probably on a large scale—especially in the Fayum oasis. Olives played an important role in the agricultural society in Karanis, as is clear from the many oil presses and olive stones that were found (Klerkx 2011). Carefully cultivated trees need to be pruned on a regular basis, which would provide the owner with a regular quantity of high-quality wood to be used as good firewood, for small objects, and for strong constructions.

Phoenix is not found in the category of the objects, but this probably has to do with its anatomy: the coarse-fibred material is not easily modified into objects. Therefore, although it is not present in every category, it is still considered to be of local origin. Cultivars are present since the second millennium BC (Zohary & Hopf 1994). The palm is valued for the fruits, and the leaves are used as roofing material and in small constructions. The long, straight, strong trunk is used in larger constructions. As firewood it is probably mainly used because it was available. Just like *Olea*, it is cultivated in plantations, but grows also in gardens and around houses.

Of these four important taxa, three are still a dominant element of the modern vegetation (Fig. 12).

Taxa only present in the objects and/or in some special constructions are *Alnus, Buxus, Cupressus, Fagus, Fraxinus,* and *Phyllyrea/Rhamnus*. These taxa were not found in the woodworking remains, the non-worked wood, or the charcoal (with the exception of some cf. *Fagus* finds), and they are not part of the native Egyptian vegetation. They are therefore considered to be imports, probably originating from the Mediterranean.

Salix, which is only present in the first two categories, could be considered to be imported. But in this case, there is one willow species that is part of the native and modern vegetation in the Fayum, namely, *Salix mucronata* Andersson (safsaf). In addition, *Salix* is also found in small amounts in the charcoal. Until there are more

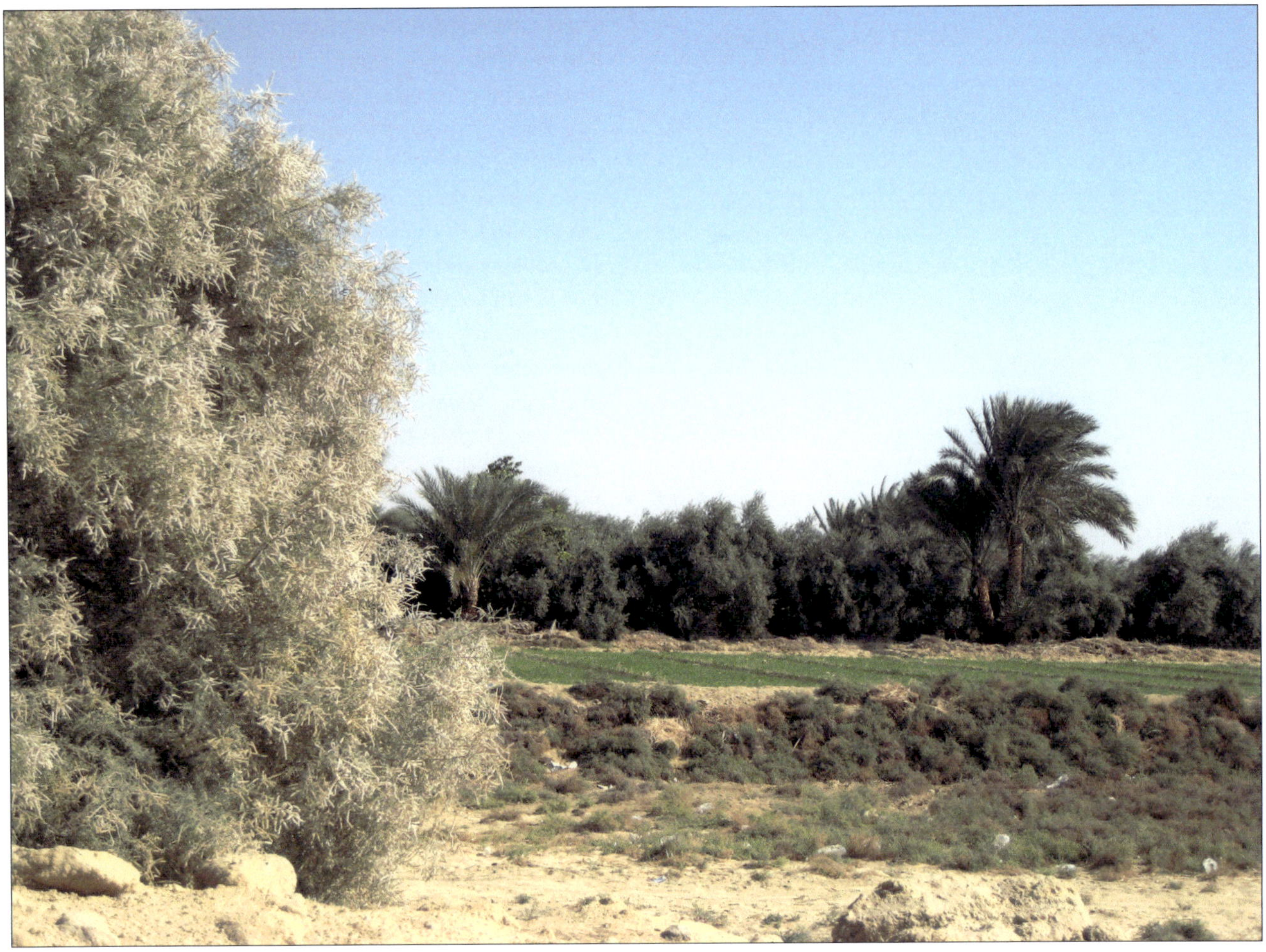

Figure 12. Modern vegetation in the Fayum oasis, showing *Phoenix*, *Olea*, and *Tamarix*.

results from categories 3 (woodworking remains) and 4 (non-worked wood), it is impossible to say whether the *Salix* finds represent imported or local trees.[5]

Abies is only found a few times, in the objects and in the woodworking remains. It cannot grow locally, so it must have been imported. It produces high-quality building material for beams and planks that may have been brought in as half-product to be finished at the site (leaving waste material: Category 3). Its absence from the construction material (Category 2) is probably due to earlier demolitions. Larger construction elements, such as floors, ceilings, and doors, are missing from the spectrum. It is possible that in these missing elements other coniferous wood taxa were present (*Cupressus, Pinus, Picea, Cedrus*), but this is difficult to prove now. *Cupressus* and *Pinus* are found as an object and in the charcoal, respectively.

There are no indications so far that any of the taxa found were chosen for symbolic or cultural reasons. Their use can be explained completely by the combination of availability and quality.

Further research, especially on material in the categories of woodworking remains and non-worked wood, will give more information on wood use in Roman Karanis.

Acknowledgments

Many thanks to René Cappers (Rijksuniversiteit Groningen) and Willeke Wendrich (University of California, Los Angeles) for making this research possible and to Laura Klerkx for her company and assistance during the field work. Welmoed Out, Kirsti Hänninen, and Erica van Hees helped while I was writing this paper. The efforts of Ursula Thanheiser in organizing the IWAA conference, where these first results were presented, deserve grateful thanks, as does the feedback provided by the reviewers.

5 The different *Salix* species cannot be differentiated based on wood anatomy.

References

Asensi Amoros V. & Détienne P., 2008. Les bois utilisés pour les portraits en Égypte à l'époque romaine. In: Aubert M.F., Cortopassi R., Nachtergael G., Asensi Amoros V., Détienne P., Pagès-Camagna S. & Le Hô A.S., 2008. *Portraits funéraires de l'Égypte romaine : cartonnages, linceuls et bois*. Éditions Khéops/Musée du Louvre éditions, Paris, pp. 29-40.

Bakels C.C., Kok R.S., Kooistra L.I. & Vermeeren C., 2000. The plant remains from Gouda-Oostpolder, a twelfth century farm in the peatlands of Holland. *Vegetation History and Archaeobotany* 9, pp. 147-160.

Boak A.E.R. & Peterson E.E., 1931. *Karanis, Topographical and Architectural Report of Excavations during the Seasons 1924–1928*. University of Michigan Press, Ann Arbor.

Boulos L., 1999. *Flora of Egypt. Vol. 1. Azollaceae–Oxalidaceae*. Al Hadara Publishing, Cairo.

Boulos L., 2000. *Flora of Egypt. Vol. 2. Geraniaceae–Boraginaceae*. Al Hadara Publishing, Cairo.

Boulos L., 2003. *Flora of Egypt. Vol. 3. Verbenaceae–Compositae*. Al Hadara Publishing, Cairo.

Boulos L., 2005. *Flora of Egypt. Vol. 4. Monocotyledons (Alismataceae–Orchidaceae)*. Al Hadara Publishing, Cairo.

Bult E.J., Hallewas D.P. & van Rijn P., 1989. Het praetorium van Valkenburg. *Natuur en Techniek* 57, pp. 194-205.

Casparie W.A., 2005. Opening up the peat bogs: The timber trackways of Drenthe. In: Louwe Kooijmans L.P., van den Broeke P.W., Fokkens H. & van Gijn A.L. (eds.), *The Prehistory of the Netherlands. Vol. I*. Amsterdam University Press, Amsterdam, pp. 401-405.

Derks T. & Vos W., 2010. Wooden combs from the Roman fort at Vechten: The bodily appearance of soldiers. *Journal of Archaeology in the Low Countries* 2 (2), pp. 53-77.

Fahn A., Werker E. & Baas P., 1986. *Wood Anatomy and Identification of Trees and Shrubs from Israel and Adjacent Regions*. Israel Academy of Sciences and Humanities, Jerusalem.

Hänninen K., 2011. Hout. In: Eijskoot Y., Brinkkemper O. & de Ridder T. (eds.), *Vlaardingen-De Vergulde Hand-West*. Rapportage Archeologische Monumentenzorg 200. Rijksdienst voor het Cultureel Erfgoed, Amersfoort, pp. 357-384.

Hänninen K. & Vermeeren C., 1995. *Giant Junipers: Houtonderzoek aan een Middenneolitsche Opgraving te Wateringen*. BIAXiaal 8. BIAX Consult, Amsterdam.

Hänninen K. & Vermeeren C., 1998. *Hout en Houtskool van een Donk met Bewoningssporen uit het Midden Neolithicum in Brandwijk (Z.H.)*. BIAXiaal 61. BIAX Consult, Amsterdam.

Hänninen K. & Vermeeren C., in prep. The use of wood: Availability or quality selection? In: Van Gijn A.L. (ed.), *Brandwijk* [working title].

Husselman E.M., 1979. *Karanis Excavations of the University of Michigan in Egypt 1928–1935: Topography and Architecture: A Summary of the Reports of the Director, Enoch E. Peterson*. University of Michigan Press, Ann Arbor.

InsideWood. 2004–present. http://insidewood.lib.ncsu.edu/search [2013].

Kalkman C., Nauta M.M. & van der Meijden R., 2003. *Planten voor Dagelijks Gebruik: Botanische Achtergronden en Toepassingen*. KNNV, Utrecht.

Klerkx L., 2011. *Archaeobotanical Investigations of Refuse Layers from Graeco-Roman Karanis (Egypt): Optimising the Interpretation of Plant Remains from Mixed Deposits*. Unpublished Masters Thesis, University of Leiden, Leiden.

Lange S., 2009. Wood analysis should include colour. In: Therkorn L.L., Besselsen E., Diepeveen-Jansen M., Gerritsen S., Karsemaken J., Kok M., Kubiak-Martens L., Slopsma J. & Vos P. (eds.), *Landcapes in the Broekpolder: Excavations around a Monument with Aspects of the Bronze Age to the Modern (Beverwijk & Heemskerk, Noord-Holland)*. Themata 2. AAC Project Bureau, Amsterdam, pp. 153-154.

Louwe Kooimans L.P., Hänninen K. & Vermeeren C., 2001. Artefacten van hout. In: Louwe Kooimans L.P. (ed.), *Hardinxveld-Giessendam De Bruin: Een Kampplaats uit het Laat-Mesolithicum en het Begin van de Swifterbant-Cultuur (5500–4450 v. Chr.)*. Rapportage Archeologische Monumentenzorg 88. Rijksdienst voor het Cultureel Erfgoed, Amersfoort, pp. 435-477.

Louwe Kooimans L.P., Vermeeren C. & van Waveren A.M.I., 2001. Artefacten van hout en vezels. In: Louwe Kooimans L.P. (ed.), *Hardinxveld-Giessendam Polderweg: Een Mesolithisch Jachtkamp in het Rivierengebied (5500–5000 v. Chr.)*. Rapportage Archeologische Monumentenzorg 83. Rijksdienst voor het Cultureel Erfgoed, Amersfoort, pp. 379-418.

Marston J.M., 2007. Report on Karanis charcoal samples excavated in 2007. Internal report Karanis Project.

Neumann K., 1989. Zur Vegetationsgeschichte der Ostsahara im Holozän: Holzkohlen aus prähistorischen Fundstellen. In: Kuper R. (ed.), *Forschungen zur Umweltgeschichte der Ostsahara*. Heinrich-Barth-Institut, Köln, pp. 13-181.

Neumann K., Schoch W., Détienne P. & Schweingruber F.H., 2000. *Woods of the Sahara and the Sahel. Bois du Sahara et du Sahel. Hölzer der Sahara und des Sahel*. Birmendorf, Eidg. Forschungsanstalt WSL, Bern, Stuttgart, and Wien.

Newton C., Terral J.-F. & Ivorra S., 2006. The Egyptian olive (*Olea europaea* subsp. *europaea*) in the later first millennium BC: Origins and history using the morphometric analysis of olive stones. *Antiquity* 80, pp. 405-414.

Oduor N.M., Wairimu Ngugi W. & wa Gathui T., 2012. *Sustainable Tree Management for Charcoal Production: Acacia Species in Kenya*. Practical Action Consulting East Africa. http://practicalaction.org/media/preview/31871.

Out W.A., 2009. *Sowing the Seed? Human Impact and Plant Subsistence in Dutch Wetlands during the Late Mesolithic and Early and Middle Neolithic (5500–3400 cal BC)*. Ph.D. Thesis. University of Leiden. Archaeological Studies Leiden University 18. Leiden.

Out W.A., 2010. Firewood collection strategies at wetland sites in the process of Neolithisation. *The Holocene* 20, pp. 191-204.

Schweingruber F.H., 1978. *Microscopic Wood Anatomy.* Swiss Federal Institute of Forestry Research, Birmensdorf.

Schweingruber F.H., 1990. *Anatomie Europäischer Hölzer. Anatomy of European woods.* Kessel, Bern and Stuttgart.

Täckholm V., Täckholm G. & Draar M., 1941. *Flora of Egypt: Vol. 1.* Bulletin of the Faculty of Science 17. Fouad I University Press, Cairo.

Täckholm V., 1974. *Students' Flora of Egypt.* 2nd ed. Cairo University Press, Beirut.

Therkorn L.L., Besselsen E., Diepeveen-Jansen M., Gerritsen S., Karsemaken J., Kok M., Kubiak-Martens L., Slopsma J. & Vos P., 2009. *Landcapes in the Broekpolder: Excavations around a Monument with Aspects of the Bronze Age to the Modern (Beverwijk & Heemskerk, Noord-Holland).* Themata 2. AAC Project Bureau, Amsterdam.

Vermeeren C., 1998. Desiccated wood and charcoal. In: Sidebotham S.E. & Wendrich W.Z. (eds.), *Berenike 1996: Preliminary Report of the Excavations at Berenike (Egyptian Red Sea Coast) and the Survey of the Eastern Desert.* Research School CNWS, Leiden, pp. 331-348.

Vermeeren C., 1999. Lady in the desert: The wood and charcoal. In: Sidebotham S.E. & Wendrich W.Z. (eds.), *Report of the 1997 Excavations at Berenike and the Survey of the Egyptian Eastern Desert, including Excavations at Shenshef.* Research School CNWS, Leiden, pp. 307-324, 427-429.

Vermeeren C., 2000/2003. *Boats in the Desert: The Desiccated Wood and Charcoal from Berenike (1998).* BIAXiaal 90. BIAX Consult, Zaandam.

Vermeeren C., 2011. Houtbeheer. In: Eijskoot Y., Brinkkemper O. & de Ridder T. (eds.), *Vlaardingen-De Vergulde Hand-West.* Rapportage Archeologische Monumentenzorg 200. Rijksdienst voor het Cultureel Erfgoed, Amersfoort, pp. 385-396.

Vermeeren C., in prep a. Wood. In: Cappers R.T.J. (ed.), *Archaeobotanical Results of Karanis* [working title].

Vermeeren C., in prep b. Wood in Roman Karanis. In: Wendrich W.Z. (ed.), *Graeco-Roman Karanis* [working title].

Vermeeren C. & Brinkkemper O., 2005. Oak or alder? The use of wood in Iron Age farms. In: Louwe Kooijmans L.P., van den Broeke P.W., Fokkens H. & van Gijn A.L. (eds.), *The Prehistory of the Netherlands. Vol. II.* Amsterdam University Press, Amsterdam, pp. 577-580.

Vermeeren C. & Cappers R.T.J., 1997/2002. *Ethnographic and Archaeobotanical Evidence of Local Cultivation of Plants in Roman Berenike and Shenshef (Red Sea Coast, Egypt).* BIAXiaal 140. BIAX Consult, Zaandam.

Vermeeren C. & De Hingh A., 2000. *Houtskoolonderzoek aan Monsters uit de Lineaire Bandkeramiek van de Opgraving Geleen-Janskamperveld 1991.* BIAXiaal 107. BIAX Consult, Amsterdam.

Wheeler E.A., Baas P. & Gasson P.E., 1989. IAWA list of microscopic features for hardwood identification. *IAWA Bulletin n.s.* 10 (3), pp. 219-332.

Wiselius S.I., 1990. *Houtvademecum.* Kluwer Technische Boeken, Deventer and Antwerp.

Zohary D. & Hopf M., 1994. *Domestication of Plants in the Old World.* 2nd ed. Oxford University Press, Oxford.